Current Topics in Microbiology and Immunology

Volume 319

Series Editors

Richard W. Compans
Emory University School of Medicine, Department of Microbiology and
Immunology, 3001 Rollins Research Center, Atlanta, GA 30322, USA

Max D. Cooper
Howard Hughes Medical Institute, 378 Wallace Tumor Institute, 1824 Sixth
Avenue South, Birmingham, AL 35294-3300, USA

Tasuku Honjo
Department of Medical Chemistry, Kyoto University, Faculty of Medicine,
Yoshida, Sakyo-ku, Kyoto 606-8501, Japan

Hilary Koprowski
Thomas Jefferson University, Department of Cancer Biology, Biotechnology
Foundation Laboratories, 1020 Locust Street, Suite M85 JAH, Philadelphia,
PA 19107-6799, USA

Fritz Melchers
Biozentrum, Department of Cell Biology, University of Basel, Klingelbergstr.
50 – 70, 4056 Basel Switzerland

Michael B.A. Oldstone
Department of Neuropharmacology, Division of Virology, The Scripps Research
Institute, 10550 N. Torrey Pines, La Jolla, CA 92037, USA

Sjur Olsnes
Department of Biochemistry, Institute for Cancer Research, The Norwegian
Radium Hospital, Montebello 0310 Oslo, Norway

Peter K. Vogt
The Scripps Research Institute, Dept. of Molecular & Exp. Medicine, Division of
Oncovirology, 10550 N. Torrey Pines. BCC-239, La Jolla, CA 92037, USA

Tim Manser
Editor

Specialization and Complementation of Humoral Immune Responses to Infection

 Springer

Tim Manser, PhD
Department of Microbiology and Immunology
Jefferson Medical College
Philadelphia, PA
USA
e-mail: manser@mail.jci.tju.edu

Background Legend: Relapsing fever Borrelia spirochetes were fixed in methanol and exposed to a fluorescent labeled a single chain Fv derived from a complement-independent bactericidal antibody to observe binding via fluorescence microscopy. This image is at 100X magnification.

Inset Legend: A negative-stained transmission electron micrograph image of a live relapsing fever Borrelia that was exposed to a single chain Fv derived from a complement-independent bactericidal IgM is shown. Membrane blebbing can be observed as well as a disruption of the normal spiral morphology of the spirochete. Exposure of periplasmic flagella is very apparent in the image and signifies that the outer membrane of the spirochete has been disrupted. The magnification of the image is 23000X.

Both images are courtesy of Tim LaRocca and Jorge Benach (see this volume)

ISBN 978-3-540-73899-2   e-ISBN 978-3-540-73900-5

Current Topics in Microbiology and Immunology ISSN 007-0217x

Library of Congress Catalog Number: 72-152360

© 2008 Springer-Verlag Berlin Heidelberg

This work is subject to copyright. All rights reserved, whether the whole or part of the material is concerned, specifically the rights of translation, reprinting, reuse of illustrations, recitation, broadcasting, reproduction on microfilm or in any other way, and storage in data banks. Duplication of this publication or parts thereof is permitted only under the provisions of the German Copyright Law of September, 9, 1965, in its current version, and permission for use must always be obtained from Springer-Verlag. Violations are liable for prosecution under the German Copyright Law.

The use of general descriptive names, registered names, trademarks, etc. in this publication does not imply, even in the absence of a specific statement, that such names are exempt from the relevant protective laws and regulations and therefore free for general use.

Product liability: The publisher cannot guarantee the accuracy of any information about dosage and application contained in this book. In every individual case the user must check such information by consulting the relevant literature.

Cover Design: WMXDesign GmbH, Heidelberg, Germany

Printed on acid-free paper

9 8 7 6 5 4 3 2 1

springer.com

Preface

The importance of specific antibodies for the clearance of and long-term resistance to many infectious pathogens has long been appreciated. Moreover, the role in these processes of the different antibody heavy chain isotypes, each tailored to induce diverse effector pathways such as those mediated by complement and Fc receptors and to promote antibody localization in distinct regions of the body, is well established. Insights into the molecular mechanism of isotype class switching showed that during an immune response B cells could change the heavy chain of the antibody they produced, without influencing the structure and specificity of the antigen-binding variable regions of this antibody (Honjo 1983). More recently, the germinal center pathway of B cell development was elucidated, in which antibody variable regions undergo extensive structural alteration via hypermutation followed by stringent phenotypic selection (Berek 1992; Kelsoe 1995). Emerging from this pathway are memory B cells and long-lived antibody-forming cells that express antigen receptors and secrete antibodies, respectively, with increased affinity and specificity for the foreign antigen.

Taken together, these findings led to a view of the acquisition of antibody-mediated resistance to infectious pathogens in mammals that involved extensive somatic maturation of antibody heavy and variable region structure and function during the immune response. This was in keeping with the concept that the B cell compartment comprises a major arm of the adaptive immune system. Such a view was reinforced by molecular analyses of antiviral antibody responses (Zinkernagel 1996). However, other past studies of antibody responses, in particular those using multivalent polysaccharide-based antigens, indicated that this notion of immunity might not apply to all pathogens. These latter examinations revealed that, unlike the antibody responses to most viruses or model protein antigens, the response to antigens derived from or thought to mimic bacterial capsule, cell wall, and outer membrane components did not require that B cells receive costimulatory signals from T cells (Mond et al. 1995). Further analysis of antibody responses to these T cell-independent (TI) antigens suggested that many did not display the same maturational characteristics as those to antigens for which efficient responses necessitated the participation of T cells (T cell-dependent, TD, antigens). In addition, investigators using hapten-protein and purified protein TD model antigens, such as myself, harbored the secret that such antigens had to be mixed with inflammation-inducing adjuvants to elicit strong antibody responses.

Superimposed on these analyses of TD and TI antibody responses were studies of the role of components of the evolutionarily hard-wired or innate immune system in conferring resistance to infectious pathogens, and of the ontogeny and function of various primary B cell subsets distinguished by cell surface phenotype and microenvironmental locale (Herzenberg and Herzenberg 1989). In the last 5 years, data from of all of these areas of research have coalesced, resulting in the emergence of a new and more complete understanding of how antibody-mediated resistance to pathogens is elaborated. The recent explosion of knowledge of Toll-like receptor (TLR) specificity and function (Takeda et al. 2003) has further embellished this understanding. It is now clear that there is not only extensive overlap and cross-complementarity in the action of innate and adaptive systems, but also specialization of function of the various B cell subsets and the types of antibodies they produce. This synergistic interaction of multiple components of these systems is perhaps best exemplified in antibody responses to bacteria. In extending invitations to potential contributors to this volume, I attempted to gather manuscripts that would highlight this new perspective on antibody responses to infection, as well as to convey its practical implications, such as for contemporary vaccine design.

The first contribution by Hinton, Jegerlehner, and Bachmann illustrates how the distinction between the function of TI and TD B cell responses in mediating resistance to viral infection has been blurred. These authors argue that patterns unique to antigens displayed on the surface of viruses and other pathogens lead to efficient TI B cell stimulation. Combined with an appreciation of the contribution of TLR recognition of viral components and the ability of TLR signaling to qualitatively and quantitatively influence the outcome of the antibody response, these findings have important implications for the design of viral vaccines and other antibody-mediated therapeutic strategies. Mond and Kokai-Kun extend this theme in their discussion of the nature of immune responses to TI antigens and the application of this knowledge to the development of vaccines for encapsulated bacteria. The idea here is that most pure TI antigens make poor pediatric vaccines—only by conjugating them to TD antigens, allowing induction of adaptive components of the immune system, is partial efficacy obtained. These authors also provide a summary of the current state of the art of vaccines for several common human bacterial pathogens, which aptly points out the inherent difficulties in translating basic discoveries in this field to practical application. Baumgarth and colleagues then discuss the role of various B cell subsets in the immune responses to viruses and bacteria, with an emphasis on B1 B cells. Their studies and those of others indicate that B1 and marginal zone B cells operate at the interface of innate and adaptive immunity. These subpopulations also cooperatively interact in the immune response to bacteria, and help in priming the more adaptive follicular B cell compartment for germinal center reactions to bacterial protein antigens and viruses.

Chapters in the second half of the volume focus on application of the concepts introduced in the first half to an understanding of how antibody responses are elicited and mediate resistance to two bacterial pathogens with very different lifestyles: the Borrelia spirochetes, causative agents of Lyme disease and relapsing fever in humans, and Streptococcus mutans, the species centrally implicated in dental caries. LaRocca and Benach provide a comprehensive depiction of the humoral factors responsible for resistance to various Borrelia subspecies, as well as associated pathologies, and also discuss immune evasion strategies of these bacteria. They also describe an intriguing class of anti-Borrelia antibodies that are bactericidal in the absence of complement. Alugupalli discusses the recently discovered ability of the B1b B cell subset to confer long-lasting, IgM-mediated resistance to Borrelia hermsii, the blood-colonizing bacterium that is a causative agent of relapsing fever in rodents. The data he has obtained indicate that B1b cells may establish a T cell-independent, TLR-dependent state of immunity to B. hermsii and other bacteria that bears many of the hallmarks of follicular B cell memory. Finally, Smith and Mattos-Graner describe the role of antibodies, the IgA isotype in particular, in controlling colonization of the tooth surface by S. mutans-containing biofilms. Their contribution nicely illustrates the often extensive and dynamic interaction of host and pathogen, and how subtle changes in the host's response to infection can lead to resistance or disease.

Philadelphia
May 2007

Tim Manser

References

Berek C (1992) The development of B cells and the B-cell repertoire in the microenvironment of the germinal center. Immunol Rev 126:5–19
Herzenberg LA, Herzenberg LA (1989) Toward a layered immune system. Cell 59:953–954
Honjo T (1983) Immunoglobulin genes. Annu Rev Immunol 1:499–528
Kelsoe G (1995) In situ studies of the germinal center reaction. Adv Immunol 60:267–288
Mond JJ, Vos Q, Lees A, Snapper CM (1995) T cell independent antigens. Curr Opin Immunol 7:349–354
Takeda K, Kaisho T, Akira S (2003) Toll-like receptors. Annu Rev Immunol 21:335–376
Zinkernagel RM (1996) Immunology taught by viruses. Science 271:173–178

Contents

**Pattern Recognition by B Cells: The Role of Antigen
Repetitiveness Versus Toll-Like Receptors** 1
H. J. Hinton, A. Jegerlehner, and M. F. Bachmann

**The Multifunctional Role of Antibodies in the Protective
Response to Bacterial T Cell-Independent Antigens** 17
J. J. Mond and J. F. Kokai-Kun

B Cell Lineage Contributions to Antiviral Host Responses 41
N. Baumgarth, Y. S. Choi, K. Rothaeusler, Y. Yang,
and L. A. Herzenberg

**The Important and Diverse Roles of Antibodies in the
Host Response to *Borrelia* Infections**............................. 63
T. J. LaRocca and J. L. Benach

**A Distinct Role for B1b Lymphocytes in
T Cell-Independent Immunity** 105
K. R. Alugupalli

**Secretory Immunity Following Mutans Streptococcal
Infection or Immunization** 131
D. J. Smith and R. O. Mattos-Graner

Index 157

Contributors

K. R. Alugupalli
Department of Microbiology and Immunology, Kimmel Cancer Center, Thomas Jefferson University, 233 South 10th Street, BLSB 726, Philadelphia, PA 19107, USA, kishore.alugupalli@mail.jci.tju.edu

M. F. Bachmann
Cytos Biotechnology AG, Wagistrasse 25, 8952 Zürich-Schlieren, Switzerland, martin.bachmann@cytos.com

N. Baumgarth
Center for Comparative Medicine, University of California, Davis, County Rd 98 & Hutchison Drive, Davis, CA 95616, USA, nbaumgarth@ucdavis.edu

J. L. Benach
Center for Infectious Diseases, 5120 Centers for Molecular Medicine, Stony Brook, NY 11794-5120, USA, jbenach@notes.cc.sunysb.edu

Y. S. Choi
Center for Comparative Medicine, University of California, Davis, County Rd 98 & Hutchison Drive, Davis, CA 95616, USA

L. A. Herzenberg
Department of Genetics, Stanford Medical School, Beckmann Center, Pasteur Drive, Stanford, CA 94205, USA

H. J. Hinton
Cytos Biotechnology AG, Wagistrasse 25, 8952 Zürich-Schlieren, Switzerland

A. Jegerlehner
Cytos Biotechnology AG, Wagistrasse 25, 8952 Zürich-Schlieren, Switzerland

J. F. Kokai-Kun
Biosynexus Incorporated, 9298 Gaither Rd, Gaithersburg, MD 20877, USA, johnkun@biosynexus.com

T. J. LaRocca
Center for Infectious Diseases, 5120 Centers for Molecular Medicine, Stony Brook, NY 11794-5120, USA

R. O. Mattos-Graner
Department of Oral Diagnosis, Piracicaba Dental School, State University of Campinas, Piracicaba, SP, Brazil

J. J. Mond
Biosynexus Incorporated, 9298 Gaither Rd, Gaithersburg, MD 20877, USA

K. Rothaeusler
Center for Comparative Medicine, University of California, Davis, County Rd 98 & Hutchison Drive, Davis, CA 95616, USA

D. J. Smith
Department of Immunology, The Forsyth Institute, 140 The Fenway, Boston, MA 02115, USA, dsmith@forsyth.org

Y. Yang
Department of Genetics, Stanford Medical School, Beckmann Center, Pasteur Drive, Stanford, CA 94205, USA

Pattern Recognition by B Cells: The Role of Antigen Repetitiveness Versus Toll-Like Receptors

H. J. Hinton, A. Jegerlehner, and M. F. Bachmann(✉)

1 Introduction: Viruses Induce Excellent Antibody Responses 2
2 Epitope Density and the Requirement for T Help 2
3 Repetitiveness as a Marker for Foreignness. 4
4 Repetitiveness to Break Self-Unresponsiveness: Practical Applications. 5
5 How Does Antigen Repetitiveness Relate to the Magnitude of the Response?. 5
6 Epitope Density and the Human B Cell Response. 7
7 Complement Activation and the BCR Signalling Threshold. 7
8 TLR Ligands and Class Switching 9
9 TLR Ligands and Vaccination 10
10 Conclusions: Implications for Vaccine Design 11
References 12

Abstract Viruses induce excellent antibody responses due to several intrinsic features. Their repetitive, organised structure is optimal for the activation of the B cell receptor (BCR), leading to an increased humoral response and a decreased dependence on T cell help. Viruses also trigger Toll-like receptors (TLRs), which in addition to increasing overall Ig levels, drive the switch to the IgG2a isotype. This isotype is more efficient in viral and bacterial clearance and will activate complement, which in turn lowers the threshold of BCR activation. Exploiting these characteristics in vaccine design may help us to create vaccines which are as safe as a recombinant vaccine yet still as effective as a virus in inducing B cell responses.

Abbreviations APC: Antigen-presenting cell; BCR: B cell receptor; Ig: Immunoglobulin; IL: Interleukin; IFN: Interferon; LPS: Lipopolysaccharide; RBC: Red blood cell; RNA: Ribonucleic acid; STAT: Signal transducers and activators of transcription; TD: T cell-dependent; TI: T cell-independent; TLR: Toll-like

M. F. Bachmann
Cytos Biotechnology AG, Wagistrasse 25, 8952 Zürich-Schlieren, Switzerland
e-mail: martin.bachmann@cytos.com

receptor; VLP: Virus-like particle; VSV: Vesicular stomatitis virus; VSV-G: Glycoprotein of vesicular stomatitis virus: XID: X-linked immunodeficiency

1 Introduction: Viruses Induce Excellent Antibody Responses

In the eighteenth century, Edward Jenner took advantage of the similarities between cowpox and smallpox to develop the first vaccine to protect against smallpox. Since then, vaccination has been used to induce antibodies against many pathogens and has proven to be an effective mechanism to protect against infectious diseases. Historically, most vaccines are either attenuated or inactivated forms of the infecting pathogens, for example the Sabin or Salk vaccines, respectively (Melnick 1996). While attenuated or inactivated vaccines are usually safe, there remains a small but real risk of reversion to a virulent phenotype. For example, vaccination against poliomyelitis has in effect eradicated the disease from the Western world. In the USA, all reported cases of poliomyelitis since 1982 have resulted from vaccination (MMWR 1986). Given the decreased probability of contracting an infectious disease owing to the success of immunisation programmes, the general public have become less accepting of the risks associated with immunisation. Therefore it has become more important to develop recombinant vaccines which are still able to induce potent and long-lasting neutralising antibody and memory B cell responses to protect against the pathogen, but which are safer and unable to revert back to virulent replicating phenotypes.

The advantage of a virus-based vaccine is that viruses themselves are able to induce strong cytotoxic T and B cell responses, while isolated antigens usually require the nonspecific inflammatory stimuli of an adjuvant to induce a protective immune response (Bachmann et al. 1998). There are a number of features which enable viruses to induce these strong responses. Firstly, viruses present epitopes in a highly structured repetitive array, which allows efficient cross-linking of the BCR. Viruses are also able to activate TLRs and complement, which, in addition to engaging the innate immune system, directly act on B cells. An additional important feature of viruses is their ability to replicate and therefore prolong the exposure to the antigen.

The challenge of modern vaccine design is to use these characteristics of viruses in the design of novel recombinant vaccines to improve their efficacy while retaining the safety of a recombinant vaccine. In this review, we will discuss these characteristics of viruses and explore how they can be used to produce improved vaccines.

2 Epitope Density and the Requirement for T Help

The induction of efficient B cell responses requires two sets of signals: the first is mediated by the BCR, while the second is usually provided by T cells (McHeyzer-Williams and McHeyzer-Williams 2005; Parker 1993). Low doses of antigen or haptenated proteins only induce B cell responses in the presence of specific T cell help. These antigens are known as T dependent (TD) (Parker 1993). In contrast, repetitive antigens such as haptenated synthetic polymers or polyclonal B cell stimuli such as LPS do not require secondary signals to induce B cell responsiveness. They are T independent (TI) and activate the B cell by extensive cross-linking of the BCR or through the activation of TLRs. Accordingly, TI antigens can be classified into type 1 (TI-1) or type 2 (TI-2) antigens. Generally TI-1 antigens are more potent B cell stimulators and can induce B cells directly with no need for antigen-presenting cells (APCs) or residual T help, while TI-2 antigens require the presence of APCs and some residual T help. TI-1 antigens tend to be polyclonal B cell activators such as LPS; in contrast TI-2 antigens are usually repetitive antigens such as haptenated polymers (Jeurissen et al. 2004; Mond et al. 1995; Vos et al. 2000; see also the chapter by Mond and Kokai-Kun, this volume).

Early studies using flagellin found that the repetitive structure of the polymerised protein is able to induce a TI B cell response, whereas monomeric flagellin or flagellin coupled to RBCs (which lacks the repetitive structure) requires T cell help to induce B cell responses (Feldmann and Basten 1971). It was later shown with haptenated polymers that a minimum number of 12–16 antigenic determinants spaced approximately 10 nm apart are able to activate B cells in the absence of T cell help (Dintzis et al. 1976). This spatially continuous cluster of antigen is coined the immunon and is the minimum immunogenic signal that is required to elicit a TI B cell response. Such responses tend to be TI-2 and do require some residual T cell help. However, highly repetitive antigens which are able to strongly cross-link the BCR are able to stimulate potent IgM responses in the complete absence of T cell help. For example, the glycoprotein of vesicular stomatitis virus (VSV-G) is not a polyclonal B cell activator but can be a TI-1 antigen. VSV has only one neutralising antigenic site, and the neutralising antibodies induced by the different forms of VSV-G all bind to the same epitope (Kelley et al. 1972; Roost et al. 1995). Therefore, it is interesting to study the effects that differences in the structure and organisation of VSV-G have on B cell responses. VSV-G is expressed on the viral envelope in a rigid, quasi-crystaline highly organised way with a spacing of 5–10 nm: the optimal structure for BCR cross-linking and subsequent B cell activation (Dintzis et al. 1976). When VSV-G is expressed on the surface of infected cells (Johnson et al. 1981), or is purified and forms micelles (Petri and Wagner 1979; Simons et al. 1978), it is motile and poorly organised. All these forms of VSV-G will induce high titres of neutralising IgM antibodies without the requirement of T help, indicating that they are TI antigens (Bachmann et al. 1995); however only the highly organised form is a TI-1 antigen. Mice with a deficiency in Bruton tyrosine kinase (BTK) such as X-linked immunodeficiency (XID) mice have immature B cells which can only be activated by TI-1 antigens. In XID mice, only highly organised VSV-G in viral particles, and not poorly organised forms, will elicit a B cell response, indicating that it is a TI-1 antigen when highly organised but not when poorly organised (Bachmann et al. 1995). This is irrespective of whether the virus is able to replicate, because formaldehyde-inactivated VSV is as efficient as WT virus at stimulating B cell responses in XID mice (Bachmann et al. 1995). Since there is only one neutralising epitope (Kelley et al. 1972; Roost et al. 1995), the differences in the B cell response cannot be attributed to differences in the epitopes recognised and must be attributed to the repetitiveness and organisation of the antigen.

These results corroborated and extended earlier experiments with haptenated beads and bacteria expressing foreign epitopes. In experiments using haptenated synthetic beads, Mond et al. found that highly haptenated beads acted as TI-1 antigens, whereas less densely haptenated beads acted as TI-2 antigens (Mond et al. 1979). Similarly, immunising mice with bacteria engineered to express foreign epitope targeted to either the cell surface (organised antigen) or the periplasmic region (less organised antigen) resulted in TI or TD responses, respectively (Leclerc et al. 1991).

3 Repetitiveness as a Marker for Foreignness

B cell responses against highly organised antigens are either independent from or less dependent on T helper cells than poorly organised or monomeric antigens. This indicates that antigen organisation is an important parameter for B cell activation. Surfaces of viruses, bacteria, and parasites tend to be repetitive, quasi-crystalline in nature. While similarly repetitive self-peptides do exist (for example actin and myosin), they tend to be localised within the cell and not accessible to B cells. In contrast, cell surface molecules almost never form stable clusters such as those found on viral surfaces but remain laterally mobile, rendering them less able to activate B cells. Additionally, membrane-bound antigens are potent inducers of B cell tolerance (Hartley et al. 1991; Russell et al. 1991). This further reduces the probability that membrane-bound antigens will be able to activate self-specific B cells, since such B cells have been deleted from the repertoire.

Together, these observations led to the hypothesis that antigen organisation (i.e. stable, quasi-crystalline surfaces as opposed to repetitive but perhaps mobile membrane proteins) may act as a marker for foreignness. This theory was tested in transgenic mice expressing the membrane form of VSV-G as a self-antigen (Bachmann et al. 1993). When these mice are immunised with poorly organised recombinant VSV-G they are unable to induce a B cell response. However, when highly organised non-replicating VSV particles are used to immunise these mice, the B cell response is restored. The absence of a B cell response cannot be attributed to tolerant T helper cells because depletion of CD4+ cells in normal mice still results in an IgM response to the recombinant VSV-G. This indicates that the IgM response is TI and the inability of disorganised VSV-G to elicit a B cell response is not caused by T cell tolerance but must result from B cell unresponsiveness. Corroborating results were found when tolerant HEL-specific B cells were stimulated in vitro with either soluble HEL or a more organised membrane form of HEL (Cooke et al. 1994). The soluble form of HEL was unable to elicit a B cell response, while a membrane-bound form was able to restore the initial signalling events involved in B cell activation. Collectively these results suggest that B cell unresponsiveness can be overcome by a highly repetitive, organised, quasi-crystalline antigen. In effect, the immune system is largely unable to distinguish between self and foreign proteins based on antigenic epitopes but does so based on antigenic organisation.

4 Repetitiveness to Break Self-Unresponsiveness: Practical Applications

The ability of highly repetitive antigens to break B cell unresponsiveness has recently been used to generate new types of vaccines for the treatment of chronic diseases. Specifically, by presenting self-peptides in an ordered array self-unresponsiveness can be overcome and autoantibodies can be raised against self-proteins. Virus-like particles (VLPs) have been developed as carriers for the induction of autoantibodies in mice (Chackerian et al. 1999, 2001, 2002; Jegerlehner et al. 2002b; Rohn et al. 2006; Spohn et al. 2005) as well as in humans (Ambuhl et al. 2007). VLPs usually consist of a recombinantly expressed viral coat protein that is assembled into particles. These particles resemble the original virus in terms of having a repetitive structure but carry no genetic information and are therefore unable to infect a host or replicate within it. The self-peptide or self-protein of interest is displayed on the surface of the VLP in a repetitive fashion, leading to a break in B cell unresponsiveness and autoantibody production. As will be discussed in Sect. 6, VLP-based vaccines, which can break B cell unresponsiveness and induce antibodies to neutralise endogenous mediators, such as inflammatory cytokines, are promising candidates for the treatment of a variety of chronic diseases (Dyer et al. 2006).

5 How Does Antigen Repetitiveness Relate to the Magnitude of the Response?

There is a wealth of evidence to suggest that in the absence of adjuvants, soluble and partially aggregated antigens induce poor IgG responses, while highly organised antigens such as bacterial (Feldmann and Basten 1971; Leclerc et al. 1991) and viral (Bachmann et al. 1995; Bachmann et al. 1993; Chackerian et al. 2001; Justewicz et al. 1995) surface proteins are able to induce strong IgG responses. Given that repetitive antigens are usually particulate, whereas poorly organised antigens are often soluble, it remains possible that the observed differences in IgG response are due to these properties rather than a consequence of antigen organisation. For example, particulate antigens are presented to T helper cells in different ways than soluble antigens and B cell responses against proteins in adjuvant are initiated at the interface of the T-B zone (Jacob and Kelsoe 1992; Liu et al. 1991; Van den Eertwegh et al. 1993), whereas particulate antigens initiate a B cell response in the marginal zone or within the B cell follicle (Bachmann et al. 1994; Gatto et al. 2004; Martin et al. 2001).

The importance of antigen repetitiveness to the magnitude of the IgG response was directly investigated using VLPs with different densities of peptide epitopes covalently attached to them (Jegerlehner et al. 2002a). In this system, VLPs with a high density of coupled peptide represent a highly organised repetitive antigen, while VLPs with a low density of coupled peptide represent a poorly organised antigen; the VLP itself provides a high-density, repetitive control epitope. The magnitude of the VLP-specific and peptide-specific IgM response was similar, irrespective of the density of the peptide on the VLP. However, while the VLP-specific IgG response was constant, the peptide-specific IgG response varied according to the density of the peptide on the VLP. High-density conjugates were able to produce a substantial IgG response while the low-density VLP-peptide conjugates were not able to produce a significant IgG response even by increasing the amount of VLP–peptide conjugate. These data confirm that antigen organisation and repetitiveness, rather than antigen concentration or its particulate structure, are critical to the overall magnitude of the IgG response. Interestingly, a similar number of epitopes as described for the immunon was found to be required for the induction of B cell responses, namely around 20 (Fig. 1).

Unexpectedly, varying the epitope density on VLPs affected the TD IgG response more than the early, TI IgM responses. Thus, the efficiency of cross-linking BCRs not

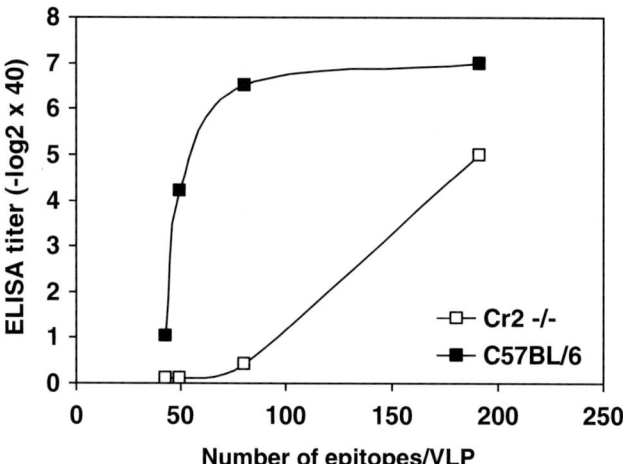

Fig. 1 Complement costimulation reduces the number of epitopes that need to be engaged by the BCR to induce an IgG response. Cr2$^{-/-}$ and C57BL/6 mice were immunised with 20 μg of VLP displaying the indicated number of peptide epitopes. The graph shows IgG titres against peptide 46 days after immunisation. The data indicate that less than 50 peptides displayed on a VLP are sufficient to induce an IgG response. Given the spherical nature of a VLP, one would expect no more than 50% of displayed peptides to be able to engage the BCR. This suggests that approximately 20 epitopes are required to elicit an IgG response – a similar number to the immunon (Dintzis et al. 1976). In the absence of co-stimulation through complement receptors, the threshold for IgG responses is shifted and a higher epitope density is required to allow efficient IgG responses

only affects IgM responses but additionally is a major driver for TD isotype-switched IgG responses. It is attractive to speculate that strong BCR cross-linking by a repetitive antigen may render the B cell more susceptible to T cell help, thereby facilitating the long-term imprint of highly repetitive antigens into the B cell repertoire.

6 Epitope Density and the Human B Cell Response

It is well known that humans mount similarly strong antibody responses to viral infections as mice and other experimental animals. Furthermore, there exists a correlation between the degree of organisation of a virus and the strength of the B cell response (Bachmann and Zinkernagel 1996). Less organised viruses induce weaker B cell responses than highly repetitive ones, the latter usually being eliminated by antibodies rather than T cells. Hence, it is likely that antigen organisation also drives human antibody responses. We have recently shown that epitopes displayed in a repetitive fashion on VLPs induce excellent antibody responses in humans even in the absence of adjuvants. As little as 10 μg of VLP displaying a house dust mite antigen is able to induce robust IgG responses in volunteers after a single injection in the absence of adjuvant (Kundig et al. 2006). Similar results are obtained when nicotine is displayed on VLPs (Maurer et al. 2005). Indeed, such a vaccine is able to help smokers to quit their habit (unpublished data). Even self-antigens displayed in a repetitive fashion on VLPs are able to induce antibody responses in humans comparable to those observed in mice and rats. Specifically, angiotensin II displayed on VLPs induces strong antibody responses in mice, rats and humans and these antibodies are able to reduce blood pressure in hypertensive rats (Ambuhl et al. 2007) and humans (unpublished data). Moreover, both ghrelin, a hormone involved in the regulation of appetite, and the proinflammatory molecule TNF induce strong IgG responses in volunteers when coupled to VLPs (unpublished data). Hence, highly repetitive antigens are able to break B cell unresponsiveness in experimental animals as well as in humans.

Such vaccines directed against self-molecules hold promise for treating a variety of chronic diseases by the induction of antibodies neutralising endogenous mediators such as inflammatory cytokines (Dyer et al. 2006). In contrast to passive vaccination with monoclonal antibodies, active immunisation with VLP-based vaccines displaying self-peptides induce long-lasting antibody responses, potentially providing affordable and convenient therapy for patients.

7 Complement Activation and the BCR Signalling Threshold

Traditionally the complement system has been viewed as a central part of the innate immune system, particularly in regards to the host defence against invading pathogens and the clearance of cell debris. However, the complement system is also important to the humoral immune response and it has been shown that deficiencies in complement components result in reduced responses to TD antigens (Bottger et al. 1986; Ellman et al. 1971; Fischer et al. 1996; Jackson et al. 1979; O'Neil et al. 1988; Ochs et al. 1983).

There are a number of mechanisms by which complement contributes to humoral immunity: complement is thought to augment the clearance of antigen to the lymphoid compartments (Fearon and Wong 1983; Ochsenbein et al. 1999); CD21–CD35 complement receptors on follicular dendritic cells enhance antigen trapping, thereby driving the germinal centre reaction and maintaining antibody titres (Barrington et al. 2002); and CD21 provides a signal independent of antigen which is required for the survival of B cells within the germinal centre (Fischer et al. 1998). An additional mechanism by which complement can regulate the B cell response is by lowering the signalling threshold of the BCR by the engagement of the CD21–CD19–CD81 complement receptor complex. Blocking the engagement of CD21 and CD35 (complement receptors 1 and 2, respectively) with either antibody or soluble receptor leads to reduced humoral responses (Gustavsson et al. 1995; Hebell et al. 1991; Heyman et al. 1990). In vitro it has also been shown that co-ligation of CD21 and the BCR lowers the threshold for B cell activation (Carter and Fearon 1992; Carter et al. 1988).

Mice deficient in the Cr2 locus which encodes CD21 and CD35 have reduced TD antibody responses (Ahearn et al. 1996). However, some antigens are able to induce potent immune responses in the absence of complement. Both VSV and influenza virus are able to induce normal responses in $Cr2^{-/-}$ mice (Kopf et al. 2002; Ochsenbein et al. 1999). As discussed above in Sect. 2, viruses possess a highly organised repetitive structure which is able to efficiently cross-link the BCR, and this feature may account for their decreased dependence on complement signals. This theory was tested using peptide epitopes coupled to VLP.

Immunisation of Cr2-deficient mice with VLPs exhibiting different epitope densities (described in Sect. 5) demonstrated that high epitope density overcame Cr2 dependence, while less repetitive antigenic forms induced Cr2-dependent responses (Jegerlehner et al. 2002a). Hence, CD19/CD21-mediated co-stimulation reduces the number of epitopes that need to be engaged (see Fig. 1). In this respect, CD19/CD21 signalling functions similarly to CD28-mediated costimulation of T cell action, where it has been shown that CD28 lowers the number of T cell receptors (TCRs) that need to be engaged for T cell activation (Bachmann et al. 1997; Viola and Lanzavecchia 1996).

Despite the activation of B cells through the complement receptors being superfluous for antibody production early after challenge with a highly repetitive antigen, it is required to maintain long-term antibody titres. In the absence of co-stimulation through the CD21 receptor complex, postgerminal centre B cells are unable to upregulate the plasma cell transcriptional regulators Blimp-1 and XBP-1, resulting in a failure to differentiate into long-lived plasma cells (Gatto et al. 2005). Thus, complement not only enhances the early B cell response, but also supports the establishment of long-term IgG memory.

8 TLR Ligands and Class Switching

TLRs belong to the family of pathogen-associated molecular pattern (PAMP) receptors which are important in distinguishing self from non-self. However, unlike the BCR which makes this distinction in an antigen-specific fashion based on the repetitive quasi-crystalline structure of pathogens, TLRs recognise small patterns or motifs commonly found in large groups of pathogens yet absent from their hosts. For example, TLR9 recognises nonmethylated DNA sequences rich in CG motifs common to micro-organisms but not vertebrates.

Like complement, TLRs have generally been thought of in terms of the induction of innate immune responses. However, recent studies have revealed the importance of TLR signalling in B cells for class switching. The balance between Th type1 (Th1) and Th type2 (Th2) is important in determining the outcome of an immune response to antigen and a response that is dominated by Th1 results in the production of the cytokine INF-γ and leads to predominantly IgG2a antibodies. Conversely, a Th2 response results in the production of the cytokines IL-4, IL-5, IL-10 and IL-13 and the production predominantly of IgG1 and IgE antibodies. Thus antigens which induce either a Th1 or Th2 response will result in the production of IgG2a or IgG1 antibodies, respectively. Similarly, in vitro experiments show that IFN-γ favours the production of IgG2a, while IL-4 favours IgG1 production. These data led to the view that Th cell-derived cytokines direct isotype switching (Coffman et al. 1988).

Some of the molecular pathways regulating IL-4 and INF-γ-induced class switching have been unravelled. The IL-4 mediated activation of STAT6 drives the switch to IgE and IgG1 through the binding and transactivation of germ-line Cγ and Cγ1 promoters (Bacharier and Geha 2000; Wurster et al. 2000). INF-γ can counteract the actions of IL-4 and drive the switch to IgG2a through the activation of STAT1 and the T-box transcription factor T-bet (Peng et al. 2002).

However, after infection with viruses and parasites, IFN-γ-independent class switching to IgG2a has been observed (Markine-Goriaynoff et al. 2000). The mechanism of this is unknown but in recent years a direct role of TLR signalling in B cells has been implicated in class switching. The addition of CpGs to B cells cultured in vitro with the Th2 cytokine IL-4 and anti-CD40L inhibited the production of IgG1 and IgE antibodies and induced the production of IgG2a, IgG2b and IgG3 (Lin et al. 2003, 2004). This correlated with an increase in expression of the transcription factor T-bet, which was dependent on TLR9, but independent of the Th1 cytokine IFN-γ and STAT1 (Liu et al. 2003). Using T-bet knockout mice, it was confirmed that T-bet is required for the CpG-induced expression of IgG2a but not for the CpG-induced inhibition of IgE/IgG1 expression (Peng et al. 2003). These experiments provide in vitro evidence that Ig class switching can be driven by the direct effects of CpG acting through TLR9 on B cells. It further identifies T-bet as the transcription factor that drives the expression of IgG2a. In vivo experiments using MyD88-null mice show that the IgG2a response to OVA-LPS is dependent on TLR signalling while the IgE response is not (Pasare and Medzhitov 2005).

In vivo, we have also found that Ig isotype switching is directed by the actions of CpGs on TLR in B cells (Jegerlehner et al. 2007). Mice were immunised with either VLP alone or with VLP containing RNA or CpG to stimulate TLR7 or TLR9, respectively. As expected, all three immunisations induced potent IgG responses, with empty VLPs inducing a response dominated by IgG1, and VLPs packaged with CpG or RNA inducing an IgG2a-dominated response. This effect was INF-γ-independent and in the case of VLP-CpG, dependent on TLR9 signalling in B cells and not dendritic cells or macrophages. Thus, TLR ligands within viral particles drive IgG2a responses by directly stimulating B cells.

There has been controversy as to whether TLR signalling is essential for B cell responses in general (Gavin et al. 2006; Nemazee et al. 2006; Pasare and Medzhitov 2005). The consensus appears to be that weakly immunogenic antigens (e.g. soluble proteins) fail to induce an IgG response in the absence of TLR ligands. In contrast, more immunogenic antigens such as VLPs or proteins in adjuvants are able to induce IgG responses in the absence of TLR signalling. Thus, for infectious agents, TLR signalling affects isotype-switching rather than the overall IgG response.

9 TLR Ligands and Vaccination

TLR ligands are generally thought to enhance adaptive immune responses; many novel adjuvant formulations therefore include such substances to increase vaccine-induced immunity. This view has, however, recently been challenged since it was shown that mice may mount good antibody responses in the absence of any TLR signalling (Gavin et al. 2006; Wickelgren 2006). This may be beside the point, as it confuses the question of whether something is required for a response with the question of whether something can enhance a response. Specifically, the fact that antibody responses can be mounted in the absence of TLR signalling does not mean that the addition of TLR ligands to a vaccine preparation will not enhance the response. Indeed, there is ample evidence in experimental animals that TLR ligands enhance antibody and in particular T cell responses. Furthermore, monophosphoryl lipid A (MPL), a ligand for TLR4, is one of the most potent adjuvants currently developed for use in humans and is included in a vaccine against human papilloma virus. Similarly, it has been shown in humans that coupling of CpGs to the hepatitis B surface antigen results in a vastly superior vaccine exhibiting a higher responder rate as well as accelerated IgG responses (Sung and Lik-Yuen 2006). Finally, we have shown that inclusion of TLR9 or TLR7/8 ligands into VLPs drives isotype switching to IgG2a in the mouse and IgG1 in humans (Jegerlehner et al. 2007). These isotypes have a much higher antibacterial and antiviral protective potential than the isotypes induced in absence of TLR-signalling, underscoring the potential value of TLR ligands for vaccine development.

10 Conclusions: Implications for Vaccine Design

In modern vaccine design, the overall aim is to create vaccines which produce high amounts of neutralising antibodies yet are as safe and well tolerated as possible. While there is no question that attenuated or inactivated viruses are excellent at inducing long-lasting protective immunity, they carry a small risk of actually causing the disease which they are designed to protect against. With this in mind, it can be more helpful to dissect out the different characteristics of viruses which enable them to effectively induce neutralising antibody responses and integrate them into a recombinant vaccine.

In this context, there are three main characteristics of interest in a virus (Table 1). Firstly, viruses present antigen in a highly structured, ordered array which allows optimal activation of the BCR leading to an overall increase in immunoglobulin levels and a decreased dependence on T cell help. In addition, viral particles are efficiently processed and presented by dendritic cells due to their particulate structure. VLPs can provide an excellent safe alternative to the virus in these respects, enabling the presentation of antigens in a particulate, highly organised array leading to optimal BCR activation.

The second viral feature which is important for obtaining good antibody responses is TLR triggering, which increases the overall magnitude of the antibody response. In addition to this, TLR triggering promotes the production of

Table 1 Summary of the effects of epitope density, TLR triggering and complement activation on the humoral immune response and how these characteristics can be utilised in vaccine design

	Effect on humoral immune response	Implication for vaccine design
Epitope repetitiveness	• Enhances the overall level of Ig produced • Reduces requirement for T cell help • Marker for foreignness	• Increases overall response to vaccine • Can break self-unresponsiveness to produce auto-antibodies to neutralise endogenous mediators of chronic diseases
TLR signalling	• Enhances the overall level of Ig produced • Promotes the switch of B cells to the IgG2a isotype	• Increases overall response to vaccine • Can modulate the isotype according to the target, i.e. IgG2a isotype is better at bacterial and viral clearance
Complement activation	• Reduces the threshold for BCR signalling • Upregulation of BLIMP-1 and XBP-1 • Differentiation to long-lived plasma cells	• Increases overall response to vaccine • Enhances long-lived plasma cell production

antibodies of the IgG2a subtype, which are particularly efficient in the defence against viral infections (Coutelier et al. 1987), partly because of their activation of complement (Klaus et al. 1979) and their ability to engage FcγRIIB (Nimmerjahn and Ravetch 2005). Given that different IgG isotypes vary in their capacity to opsonise bacteria and lyse infected cells, the TLR-mediated control of isotype induction can be used to modulate the IgG isotype according to the target of the vaccine.

The third viral characteristic which is instrumental in inducing good antibody responses is replication, which acts to prolong the exposure to the antigen. However, any vaccine that is able to replicate will also potentially be able to revert back to virulence and hence will not be completely safe. As a safe alternative, adjuvants which create a depot of antigen as well as boosting regimes may prolong the exposure to antigen and increase the antibody response in the absence of antigen replication.

References

Ahearn JM, Fischer MB, Croix D, Goerg S, Ma M, Xia J, Zhou X, Howard RG, Rothstein TL, Carroll MC (1996) Disruption of the Cr2 locus results in a reduction in B-1a cells and in an impaired B cell response to T-dependent antigen Immunity 4:251–262

Ambuhl PM, Tissot AC, Fulurija A, Maurer P, Nussberger J, Sabat R, Nief V, Schellekens C, Sladko K, Roubicek K, Pfister T, Rettenbacher M, Volk HD, Wagner F, Muller P, Jennings GT, Bachmann MF (2007) A vaccine for hypertension based on virus-like particles: preclinical efficacy and phase I safety and immunogenicity J Hypertens 25:63–72

Bacharier LB, Geha RS (2000) Molecular mechanisms of IgE regulation J Allergy Clin Immunol 105: S547–S558

Bachmann MF, Rohrer UH, Kundig TM, Burki K, Hengartner H, Zinkernagel RM (1993) The influence of antigen organization on B cell responsiveness Science 262:1448–1451

Bachmann MF, Kundig TM, Odermatt B, Hengartner H, Zinkernagel RM (1994) Free recirculation of memory B cells versus antigen-dependent differentiation to antibody-forming cells. J Immunol 153:3386–3397

Bachmann MF, Hengartner H, Zinkernagel RM (1995) T helper cell-independent neutralizing B cell response against vesicular stomatitis virus: role of antigen patterns in B cell induction? Eur J Immunol 25:3445–3451

Bachmann MF, Zinkernagel RM (1996) The influence of virus structure on antibody responses and virus serotype formation. Immunol Today 17:553–558

Bachmann MF, McKall-Faienza K, Schmits R, Bouchard D, Beach J, Speiser DE, Mak TW, Ohashi PS (1997) Distinct roles for LFA-1 and CD28 during activation of naive T cells: adhesion versus costimulation. Immunity 7:549–557

Bachmann MF, Zinkernagel RM, Oxenius A (1998) Immune responses in the absence of costimulation: viruses know the trick. J Immunol 161:5791–5794

Barrington RA, Pozdnyakova O, Zafari MR, Benjamin CD, Carroll MC (2002) B lymphocyte memory: role of stromal cell complement and Fc{gamma}RIIB receptors. J Exp Med 196:1189–1200

Bottger EC, Metzger S, Bitter-Suermann D, Stevenson G, Kleindienst S, Burger R (1986) Impaired humoral immune response in complement C3-deficient guinea pigs: absence of secondary antibody response. Eur J Immunol 16:1231–1235

Carter RH, Fearon DT (1992) CD19: lowering the threshold for antigen receptor stimulation of B lymphocytes. Science 256:105–107

Carter RH, Spycher MO, Ng YC, Hoffman R, Fearon DT (1988) Synergistic interaction between complement receptor type 2 and membrane IgM on B lymphocytes. J Immunol 141:457–463

Chackerian B, Lowy DR, Schiller JT (1999) Induction of autoantibodies to mouse CCR5 with recombinant papillomavirus particles. Proc Natl Acad Sci U S A 96:2373–2378

Chackerian B, Lowy DR, Schiller JT (2001) Conjugation of a self-antigen to papillomavirus-like particles allows for efficient induction of protective autoantibodies. J Clin Invest 108:415–423

Chackerian B, Lenz P, Lowy DR, Schiller JT (2002) Determinants of autoantibody induction by conjugated papillomavirus virus-like particles. J Immunol 169:6120–6126

Coffman RL, Seymour BW, Lebman DA, Hiraki DD, Christiansen JA, Shrader B, Cherwinski HM, Savelkoul HF, Finkelman FD, Bond MW et al (1988) The role of helper T cell products in mouse B cell differentiation and isotype regulation. Immunol Rev 102:5–28

Cooke MP, Heath AW, Shokat KM, Zeng Y, Finkelman FD, Linsley PS, Howard M, Goodnow CC (1994) Immunoglobulin signal transduction guides the specificity of B cell–T cell interactions and is blocked in tolerant self-reactive B cells. J Exp Med 179:425–438

Coutelier JP, van der Logt JT, Heessen FW, Warnier G, Van Snick J (1987) IgG2a restriction of murine antibodies elicited by viral infections. J Exp Med 165:64–69

Dintzis HM, Dintzis RZ, Vogelstein B (1976) Molecular determinants of immunogenicity: the immunon model of immune response. Proc Natl Acad Sci U S A 73:3671–3675

Dyer MR, Renner WA, Bachmann MF (2006) A second vaccine revolution for the new epidemics of the 21st century. Drug Discov Today 11:1028–1033

Ellman L, Green I, Judge F, Frank MM (1971) In vivo studies in C4-deficient guinea pigs. J Exp Med 134:162–175

Fearon DT, Wong WW (1983) Complement ligand-receptor interactions that mediate biological responses. Annu Rev Immunol 1:243–271

Feldmann M, Basten A (1971) The relationship between antigenic structure and the requirement for thymus-derived cells in the immune response. J Exp Med 134:103–119

Fischer MB, Ma M, Goerg S, Zhou X, Xia J, Finco O, Han S, Kelsoe G, Howard RG, Rothstein TL, Kremmer E, Rosen FS, Carroll MC (1996) Regulation of the B cell response to T-dependent antigens by classical pathway complement. J Immunol 157:549–556

Fischer MB, Goerg S, Shen L, Prodeus AP, Goodnow CC, Kelsoe G, Carroll MC (1998) Dependence of germinal center B cells on expression of CD21/CD35 for survival. Science 280:582–585

Gatto D, Ruedl C, Odermatt B, Bachmann MF (2004) Rapid response of marginal zone B cells to viral particles J Immunol 173:4308–4316

Gatto D, Pfister T, Jegerlehner A, Martin SW, Kopf M, Bachmann MF (2005) Complement receptors regulate differentiation of bone marrow plasma cell precursors expressing transcription factors Blimp-1 and XBP-1. J Exp Med 201:993–1005

Gavin AL, Hoebe K, Duong B, Ota T, Martin C, Beutler B, Nemazee D (2006) Adjuvant-enhanced antibody responses in the absence of toll-like receptor signaling. Science 314:1936–1938

Gustavsson S, Kinoshita T, Heyman B (1995) Antibodies to murine complement receptor 1 and 2 can inhibit the antibody response in vivo without inhibiting T helper cell induction. J Immunol 154:6524–6528

Hartley SB, Crosbie J, Brink R, Kantor AB, Basten A, Goodnow CC (1991) Elimination from peripheral lymphoid tissues of self-reactive B lymphocytes recognizing membrane-bound antigens Nature 353:765–769

Hebell T, Ahearn JM, Fearon DT (1991) Suppression of the immune response by a soluble complement receptor of B lymphocytes. Science 254:102–105

Heyman B, Wiersma EJ, Kinoshita T (1990) In vivo inhibition of the antibody response by a complement receptor-specific monoclonal antibody. J Exp Med 172:665–668

Jackson CG, Ochs HD, Wedgwood RJ (1979) Immune response of a patient with deficiency of the fourth component of complement and systemic lupus erythematosus. N Engl J Med 300:1124–1129

Jacob J, Kelsoe G (1992) In situ studies of the primary immune response to (4-hydroxy-3-nitrophenyl)acetyl. II. A common clonal origin for periarteriolar lymphoid sheath-associated foci and germinal centers. J Exp Med 176:679–687

Jegerlehner A, Storni T, Lipowsky G, Schmid M, Pumpens P, Bachmann MF (2002a) Regulation of IgG antibody responses by epitope density and CD21-mediated costimulation. Eur J Immunol 32:3305–3314

Jegerlehner A, Tissot A, Lechner F, Sebbel P, Erdmann I, Kundig T, Bachi T, Storni T, Jennings G, Pumpens P, Renner WA, Bachmann MF (2002b) A molecular assembly system that renders antigens of choice highly repetitive for induction of protective B cell responses. Vaccine 20:3104–3112

Jegerlehner A, Maurer P, Bessa J, Hinton HJ, Kopf M, Bachmann MF (2007) TLR9 signaling in B cells determines class switch recombination to IgG2a. J Immunol 178:2415–2420

Jeurissen A, Ceuppens JL, Bossuyt X (2004) T lymphocyte dependence of the antibody response to 'T lymphocyte independent type 2' antigens. Immunology 111:1–7

Johnson DC, Schlesinger MJ, Elson EL (1981) Fluorescence photobleaching recovery measurements reveal differences in envelopment of Sindbis and vesicular stomatitis viruses. Cell 23:423–431

Justewicz DM, Doherty PC, Webster RG (1995) The B-cell response in lymphoid tissue of mice immunized with various antigenic forms of the influenza virus hemagglutinin. J Virol 69:5414–5421

Kelley JM, Emerson SU, Wagner RR (1972) The glycoprotein of vesicular stomatitis virus is the antigen that gives rise to and reacts with neutralizing antibody. J Virol 10:1231–1235

Klaus GG, Pepys MB, Kitajima K, Askonas BA (1979) Activation of mouse complement by different classes of mouse antibody. Immunology 38:687–695

Kopf M, Abel B, Gallimore A, Carroll M, Bachmann MF (2002) Complement component C3 promotes T-cell priming and lung migration to control acute influenza virus infection. Nat Med 8:373–378

Kundig TM, Senti G, Schnetzler G, Wolf C, Prinz Vavricka BM, Fulurija A, Hennecke F, Sladko K, Jennings GT, Bachmann MF (2006) Der p 1 peptide on virus-like particles is safe and highly immunogenic in healthy adults. J Allergy Clin Immunol 117:1470–1476

Leclerc C, Charbit A, Martineau P, Deriaud E, Hofnung M (1991) The cellular location of a foreign B cell epitope expressed by recombinant bacteria determines its T cell-independent or T cell-dependent characteristics. J Immunol 147:3545–3552

Lin L, Gerth AJ, Peng SL (2004) CpG DNA redirects class-switching towards "Th1-like" Ig isotype production via TLR9 and MyD88. Eur J Immunol 34:1483–1487

Liu N, Ohnishi N, Ni L, Akira S, Bacon KB (2003) CpG directly induces T-bet expression and inhibits IgG1 and IgE switching in B cells. Nat Immunol 4:687–693

Liu YJ, Zhang J, Lane PJ, Chan EY, MacLennan IC (1991) Sites of specific B cell activation in primary and secondary responses to T cell-dependent and T cell-independent antigens. Eur J Immunol 21:2951–2962

Markine-Goriaynoff D, van der Logt JT, Truyens C, Nguyen TD, Heessen FW, Bigaignon G, Carlier Y, Coutelier JP (2000) IFN-gamma-independent IgG2a production in mice infected with viruses and parasites. Int Immunol 12:223–230

Martin F, Oliver AM, Kearney JF (2001) Marginal zone and B1 B cells unite in the early response against T-independent blood-borne particulate antigens. Immunity 14:617–629

Maurer P, Jennings GT, Willers J, Rohner F, Lindman Y, Roubicek K, Renner WA, Muller P, Bachmann MF (2005) A therapeutic vaccine for nicotine dependence: preclinical efficacy, and phase I safety and immunogenicity. Eur J Immunol 35:2031–2040

McHeyzer-Williams LJ, McHeyzer-Williams MG (2005) Antigen-specific memory B cell development. Annu Rev Immunol 23:487–513

Melnick JL (1996) Current status of poliovirus infections. Clin Microbiol Rev 9:293–300

MMWR (1986) Update: influenza activity – United States. MMWR Morb Mortal Wkly Rep 35:249–251
Mond JJ, Lees A, Snapper CM (1995) T cell-independent antigens type 2. Annu Rev Immunol 13:655–692
Mond JJ, Stein KE, Subbarao B, Paul WE (1979) Analysis of B cell activation requirements with TNP-conjugated polyacrylamide beads. J Immunol 123:239–245
Nemazee D, Gavin A, Hoebe K, Beutler B (2006) Immunology: Toll-like receptors and antibody responses. Nature 441:E4; discussion E4
Nimmerjahn F, Ravetch JV (2005) Divergent immunoglobulin g subclass activity through selective Fc receptor binding. Science 310:1510–1512
O'Neil KM, Ochs HD, Heller SR, Cork LC, Morris JM, Winkelstein JA (1988) Role of C3 in humoral immunity. Defective antibody production in C3-deficient dogs. J Immunol 140:1939–1945
Ochs HD, Wedgwood RJ, Frank MM, Heller SR, Hosea SW (1983) The role of complement in the induction of antibody responses. Clin Exp Immunol 53:208–216
Ochsenbein AF, Pinschewer DD, Odermatt B, Carroll MC, Hengartner H, Zinkernagel RM (1999) Protective T cell-independent antiviral antibody responses are dependent on complement. J Exp Med 190:1165–1174
Parker DC (1993) T cell-dependent B cell activation. Annu Rev Immunol 11:331–360
Pasare C, Medzhitov R (2005) Control of B-cell responses by Toll-like receptors Nature 438:364–368
Peng SL, Li J, Lin L, Gerth A (2003) The role of T-bet in B cells. Nat Immunol 4:1041; author reply 1041
Peng SL, Szabo SJ, Glimcher LH (2002) T-bet regulates IgG class switching and pathogenic autoantibody production. Proc Natl Acad Sci U S A 99:5545–5550
Petri WA Jr, Wagner RR (1979) Reconstitution into liposomes of the glycoprotein of vesicular stomatitis virus by detergent dialysis. J Biol Chem 254:4313–4316
Rohn TA, Jennings GT, Hernandez M, Grest P, Beck M, Zou Y, Kopf M, Bachmann MF (2006) Vaccination against IL-17 suppresses autoimmune arthritis and encephalomyelitis. Eur J Immunol 36:2857–2867
Roost HP, Bachmann MF, Haag A, Kalinke U, Pliska V, Hengartner H, Zinkernagel RM (1995) Early high-affinity neutralizing anti-viral IgG responses without further overall improvements of affinity. Proc Natl Acad Sci U S A 92:1257–1261
Russell DM, Dembic Z, Morahan G, Miller JF, Burki K, Nemazee D (1991) Peripheral deletion of self-reactive B cells. Nature 354:308–311
Simons K, Helenius A, Leonard K, Sarvas M, Gething MJ (1978) Formation of protein micelles from amphiphilic membrane proteins. Proc Natl Acad Sci U S A 75:5306–5310
Spohn G, Schwarz K, Maurer P, Illges H, Rajasekaran N, Choi Y, Jennings GT, Bachmann MF (2005) Protection against osteoporosis by active immunization with TRANCE/RANKL displayed on virus-like particles. J Immunol 175:6211–6218
Sung JJ, Lik-Yuen H (2006) HBV-ISS (Dynavax). Curr Opin Mol Ther 8:150–155
Van den Eertwegh AJ, Noelle RJ, Roy M, Shepherd DM, Aruffo A, Ledbetter JA, Boersma WJ, Claassen E (1993) In vivo CD40-gp39 interactions are essential for thymus-dependent humoral immunity. I. In vivo expression of CD40 ligand, cytokines, and antibody production delineates sites of cognate T-B cell interactions. J Exp Med 178:1555–1565
Viola A, Lanzavecchia A (1996) T cell activation determined by T cell receptor number and tunable thresholds Science 273:104–106
Vos Q, Lees A, Wu ZQ, Snapper CM, Mond JJ (2000) B-cell activation by T-cell-independent type 2 antigens as an integral part of the humoral immune response to pathogenic microorganisms. Immunol Rev 176:154–170
Wickelgren I (2006) Immunology. Mouse studies question importance of toll-like receptors to vaccines. Science 314:1859–1860
Wurster AL, Tanaka T, Grusby MJ (2000) The biology of Stat4 and Stat6. Oncogene 19:2577–2584

The Multifunctional Role of Antibodies in the Protective Response to Bacterial T Cell-Independent Antigens

J. J. Mond and J. F. Kokai-Kun(✉)

1 Introduction 18
2 Carbohydrates as TI Antigens 19
3 Conjugate Vaccines 21
4 Immune Response to Encapsulated Bacteria. 23
5 Commercial Vaccine Efforts Targeting T Cell-Independent Antigens 25
   5.1 Haemophilus Influenzae Type B 25
   5.2 Neisseria Meningitides 29
   5.3 *Streptococcus Pneumoniae* 31
   5.4 *Staphylococcus Aureus* 32
   5.5 *Salmonella Typhi*. 35
6 Conclusion. 35
References 36

Abstract While most complex antigens can induce antibody responses in a mature immunological system, this is not the case when injected into ontogenetically immature systems, as are found in neonates and pediatric-age children. Thus the antibody response to polysaccharides, which would in theory provide protection against infection by all polysaccharide encapsulated bacteria, including *Streptococcus pneumoniae*, *Neisseria meningitides*, and *Haemophilus influenzae*, cannot be stimulated by immunization with the polysaccharides by themselves. It was only with the introduction of conjugate vaccines that protection from these bacterial infections was provided to this susceptible age group. The introduction of these conjugate vaccines into the arsenal of vaccines serves as a remarkable example of how valuable it is to understand the mechanisms of biological processes. Many years of intense laboratory investigation demonstrated that when polysaccharides are covalently conjugated to proteins, the characteristics of the immune response are similar to that of the protein rather than the polysaccharide. These characteristics would induce an anti-polysaccharide response even in the pediatric population, which was heretofore unable to mount protective responses

---

J. F. Kokai-Kun
Biosynexus Incorporated, 9298 Gaither Rd., Gaithersburg, MD 20877, USA
e-mail: johnkun@biosynexus.com

to the polysaccharide. With the advent of conjugate vaccines for the above three mentioned bacteria, the incidence of bacteremia, meningitis, and otitis media has almost been eliminated. This chapter discusses in some detail the mechanisms which underlie the effectiveness of conjugate vaccines and discusses some of the vaccines that have been commercialized.

Abbreviations TD: T cell-dependent; TI: T cell-independent; TLR: Toll-like receptor; BCR: B cell receptor; PS: Polysaccharide; MHC: Major histocompatability complex; PAMPS: Pathogen-associated molecular patterns; TNF: Tumor necrosis factor; NK: Natural killer; PRP: Polyribosyl ribitol phosphate; Hib: *Haemophilus influenzae* type b; ACIP: Advisory Committee on Immunization Practices; LTA: Lipoteichoic acid; *xid*: X-linked immune defect

1 Introduction

The nature of the adaptive immune response has been shown to be directly dependent not only on the chemical/physical properties of the immunogen but also on its ability to stimulate cytokines. Thus soluble protein antigens stimulate T cell-dependent (TD) antibody responses of multiple isotypes and of a wide range of antibody avidities. Most soluble polysaccharides, on the other hand, are unable to stimulate classical T cell help and consequently directly stimulate B cells in a T cell-independent (TI) fashion, resulting in antibody of more restricted isotypes and restricted antibody avidity (Boswell and Stein 1996; Mond et al. 1995). While the magnitude and duration of the antibody response to a TD antigen depends on the extent of T cell stimulation and the nature of the T cells stimulated, the immune response to a TI antigen depends on its ability to stimulate ancillary signals via the complement receptor (Dempsey et al. 1996; Pozdnyakova et al. 2003), Toll-like receptors (TLR), which are present on B cells and on phagocytic cells, and other innate immune receptors (Zarember and Godowski 2002). In the natural world of infectious organisms, this classification of TI and TD antigens plays an important role in allowing us to understand the steps in the adaptive immune response that lead to protection and thereby enable us to design effective vaccines tailored to the specific pathogen. In this article, we will focus on the responses to encapsulated bacteria and more specifically on responses to the polysaccharide capsule of these organisms.

The public face that a bacterial pathogen often presents to the immune system of an infected host is a polysaccharide capsule of varying thickness. This capsule has several benefits in terms of the pathogen, serving to hide the bacteria from the immune surveillance of the host. In some pathogenic bacteria such as *Neisseria* species and *Haemophilus*, the capsule serves as a primary virulence factor, while for other bacteria such as *Staphylococcus aureus*, the contribution of the capsule to virulence is still in question and this contribution appears to be related to the site and type of infection.

2 Carbohydrates as TI Antigens

T cell-independent antigens are capable of stimulating B cell proliferation and differentiation in the absence of T cell help, most likely as a result of their polyvalent nature, i.e., the presence of many identical repeating units on the molecule (Brunswick et al. 1988). While all antigens engage the B cell receptor (BCR) and induce activational events, those induced by multivalent antigens are distinctly different than those induced by paucivalent antigens (Brunswick et al. 1988; Snapper and Mond 1996; see also the chapter by H.J. Hinton et al., this volume). The outcome of engagement of the B cell receptor is complex and can result in a wide range of responses, including proliferation, differentiation, cytokine secretion, antigen presentation, memory B cell induction, and apoptosis (Niiro and Clark 2002). The higher the valency of the antigen the more likely it will induce multivalent cross-linking of the BCR and thus induce activation at lower concentrations of antigen in both mature and immature B cells (Brunswick et al. 1988). This finding was shown to reflect its ability to stabilize B cell antigen receptor surface signaling domains (Blery et al. 2006; Sproul et al. 2000). While both paucivalent and multivalent antigens induced the redistribution of the surface BCR into polarized caps, the polyvalent antigen-induced BCR caps persisted, whereas the paucivalent-induced caps dissipated. These caps or lipid rafts are microdomains which are rich in cholesterol, and many of the signaling kinases become associated with these rafts upon signaling and thus determine the extent and persistence of the downstream signaling events (Karnell et al. 2005; Sproul et al. 2000). Supporting this hypothesis is the finding that antigen receptors on anergic B cells are endocytosed at a very enhanced rate upon binding antigen. Earlier studies have shown that persistence of calcium mobilization was another factor that determined whether BCR engagement would lead to cellular activation or not (Brunswick et al. 1989). Later studies showed that BCR engagement could lead not only to cell activation but also lead to apoptosis when membrane Ig was cross-linked in the absence of other costimulatory signals to the B cells (Fuentes-Panana et al. 2004; Koedel et al. 2003; Malissein et al. 2006; Weintraub et al. 2000).

The T cell-independent antigens have been viewed not as a single class of antigens but rather as two distinct classes. One set represented by polysaccharide antigens are classified as type 2 TI antigens and the other represented by lipopolysaccharide are classified as type 1 TI antigens (Vos et al. 2000). One of the major distinctions is that the former are unable to stimulate responses in neonatal B cells and the latter is able to stimulate neonatal immune responses most likely as a result of their ability to costimulate Toll-like receptor (TLR) on B cells and on phagocytic cells, which then secrete stimulatory cytokines and provide ancillary help. These conclusions have been supported by experiments done both in vitro and in animal models (Vos et al. 2000). For the most part, however, these conclusions are derived from events that occur in the perfect world of synthetically derived antigens and not under conditions that prevail during active infection with encapsulated bacteria. While free polysaccharide may in fact be released from dying cells or dividing cells, this polysaccharide is likely associated with other bacterial derived ligands, including, for example, lipoproteins or peptidoglycans which are TLR ligands (AlonsoDeVelasco et al. 1995; Malley et al. 2003; Schwandner et al. 1999). Thus in fact, in vivo, all polysaccharide antigens released from infecting bacteria may behave like type 1 TI antigens. This has been shown to be the case in vitro by Sen et al., who demonstrated that many polysaccharide preparations have cytokine-inducing properties that are not due to the polysaccharide (PS) but rather to the molecules associated with the PS (Sen et al. 2005). While this adjuvant-like association enhances the immunogenicity of polysaccharides, it does not replace the advantages that could be provided by T cell recruitment, most importantly the generation of germinal centers, memory induction, and high avidity-antibody. It also highlights the fact that the unresponsiveness of neonatal B cells to polysaccharides does not result from a hole in the antibody repertoire of immature B cells but rather a cellular unresponsiveness which can be overcome by the introduction of ancillary signals including TLR-mediated signaling (Landers and Bondada 2005). This point had been suggested earlier, prior to the discovery of TLR, by the demonstration that the combination of antigen and B cell-activating substances could stimulate X-linked immune defect (*xid*) B cells with the Btk mutation (Ahmad and Mond 1986). These *xid* mice show marked reductions in their antibody response to soluble T cell-independent type 2 antigens such as polysaccharide antigens and thus mimic responses of neonatal B cells (Ahmad and Mond 1986; Khan et al. 1995; Mosier et al. 1977; Scher et al. 1975). Many of the mechanisms that underlie the unresponsiveness of neonatal B cells to polysaccharide antigens has been inferred from mice with the Btk mutation. Thus the unresponsiveness of both neonatal and Btk mutant mice can be overcome in a number of different ways, including insolubilizing the antigen and thereby presenting it in a particulate form (Mond et al. 1979b), addition of a TLR agonist to the antigen (Ahmad and Mond 1986) or providing T cell help (Golding et al. 1982). In support of this, it was shown that *xid* mice immunized with intact *Streptococcus pneumoniae* type 14 that can recruit both T cell help as well as stimulate cytokine production, elicited IgM and IgG anti-polysaccharide responses, albeit substantially reduced (Khan et al. 2005). Neonatal B cells similar to *xid* B cells exhibit suppressed immune responses to polysaccharides and show restoration of responses with similar approaches as used with the *xid* B cells (Landers and Bondada 2005). The neonatal defect that underlies unresponsiveness to PS antigens has been described to be at the level of the B cell as well as at the level of the cytokine-secreting phagocytic cells (Hayward and Lawton 1977). Thus the neonatal B cell is defective in upregulation of MHC class II molecules after engagement of the BCR (Mond et al. 1979a) and is more susceptible to apoptotic stimuli than the mature B cell after antigen encounter (Sater et al. 1998).

3 Conjugate Vaccines

In view of the protective effects of anti-polysaccharide antibodies to encapsulated bacteria, it was necessary to devise an approach to induce anti-polysaccharide responses in neonates who are unresponsive to these antigens, both to provide protective levels of antibody and to induce the generation of memory B cells that could respond to subsequent encounter with these bacterial antigens. There was a large body of literature that demonstrated that when small antigens (haptens) were covalently conjugated to T cell-dependent antigens (carriers), the activated T cells could enhance the response to the otherwise nonimmunogenic haptens (Mitchison 2004). Based on this seminal finding, many investigators with an interest in pediatric vaccines prepared covalent conjugates of polysaccharides and T cell-dependent carrier molecules (Donnelly et al. 1990; Kuo et al. 1995; Peeters et al. 1991; Robbins and Schneerson 1990). When mice were immunized with these polysaccharide–protein conjugates, anti-polysaccharide responses were stimulated that had the characteristics of classical T cell-dependent antigens, including immunological memory and antibody of multiple IgG isotypes. Furthermore, while the response to polysaccharides are found predominantly in the marginal zone population of B cells (Martin and Kearney 2002), these protein–polysaccharide conjugates were taken up by follicular B cells via the polysaccharide-specific B cell receptor. As shown in Fig. 1, the antigen is internalized, processed into peptides and presented in the context of B cell surface major histocompatibility complex class II molecules where they activate peptide-specific T cells (MacLennan et al. 2003; Martin and Kearney 2002). In turn, these activated T cells engage PS-specific B cells via cognate T cell–B cell interaction and engagement of costimulatory molecules on the cells, which leads to activation of the B cells, differentiation into Ig-secreting cells, germinal center formation in preparation for memory cell generation, and somatic hypermutation leading to an increase in antibody avidity. The memory B cells that are generated are poised to respond to bacterial polysaccharide antigens during a subsequent infection. While any carrier molecule that is able to stimulate T cells can be used as the carrier molecule, different carrier molecules possess different properties which affect the height of the response and the rapidity of the response. Thus, for example, carrier molecules that have inherent TLR-stimulating properties stimulate more rapid responses than those carriers which do not have TLR-stimulating properties (Donnelly et al. 1990; Latz et al. 2004). In all the conjugate vaccine preparations that are available, only a small handful of carrier proteins have been used, and this has already presented a practical problem of antigenic competition (Mawas et al. 2006). Thus, repeated use of the same carrier molecule with different polysaccharides results in a diminished response as compared to the response induced by the polysaccharide conjugated to different carrier molecules. This issue of antigenic competition is becoming of increasing concern because the number of different conjugated polysaccharides is increasing with each injection. Currently, there is an 11-valent polysaccharide–protein vaccine preparation that is being developed against *S. pneumoniae* (see Sect. 5 below for more information on commercially available conjugate vaccines).

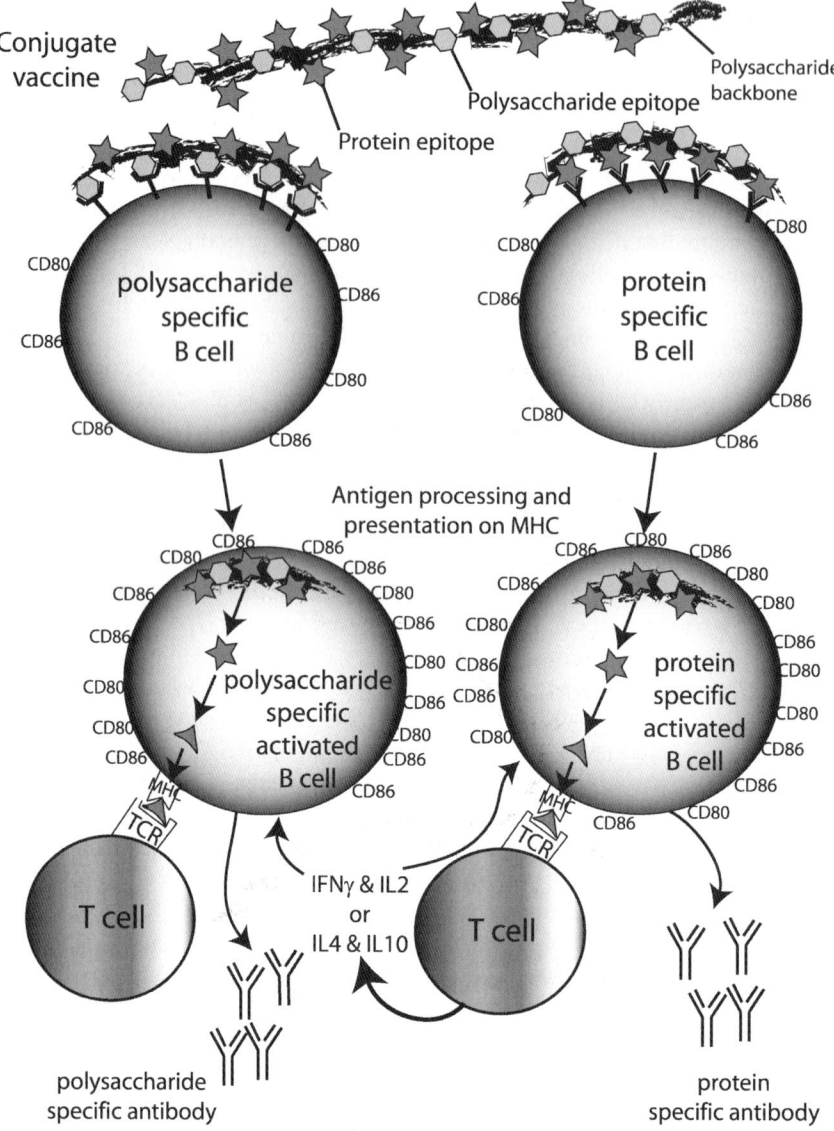

Fig. 1 Immunologic recognition of conjugate vaccines. Conjugate vaccines consisting of multiple copies of a single protein chemically linked to a polysaccharide backbone bind to the membrane immunoglobulin of B cells that recognize either the polysaccharide (PS) antigens (depicted as hexagons) or the protein determinants (depicted as stars) present on the vaccine. The PS displays multiple and identical repeating determinants and is therefore an extremely effective ligand for inducing multipoint cross-linking of the Ig on the polysaccharide-specific B cell. This cross-linking rapidly stimulates cellular activation reflected in enhanced expression of costimulatory molecules, including MHC class II, and thereby becomes a very effective antigen-presenting cell. As the conjugate vaccine is bound and internalized by this activated cell, the protein antigens are dissociated from the polysaccharide and further processed within the cell along the same pathways as are unconjugated protein antigens. Peptide fragments derived from the protein portion of the conjugate

4 Immune Response to Encapsulated Bacteria

For the most part, vaccine development has focused on the elicitation of anti-polysaccharide antibodies primarily because these antigens are so highly expressed on the surface of the organism and they are disease promoting because of their anti-opsonic activity. More recently, a number of laboratories have shown that anti-protein responses can also be protective in animal models and some of these antigens are being advanced into clinical trials (Baril et al. 2006). In designing the optimal vaccine, it is critical to understand the mechanisms underlying the host's adaptive response to the intact organism since vaccination is designed to induce memory B and/or T cells that would be rechallenged by the organism or fragments derived from the organism during infection.

The immune response to extracellular bacteria is not easily compartmentalized because it is so complex. Bacteria are particulate antigens that express many different surface proteins, polysaccharides, and potentially multiple TLR ligands. The ingestion of bacteria by cells of the phagocytic system is followed by a complex pathway of antigen processing of the different cellular components, leading to activation of different mechanisms of the adaptive immune response (Balazs et al. 2002). One of the early findings that suggested that responses to the intact bacterium were in fact different than that to the isolated components of the bacterium was the observation that there was a high degree of somatic mutation and diversification in the anti-polysaccharide response to the intact organism, which was not the case for the response to the isolated polysaccharide (Casal and Tarrago 2003). This implicates the antigen-presenting cell and the recruitment of cognate T cell help in the response. Furthermore, it helps define the subset of B cells that are recruited in the response. Thus, it is the marginal zone B cell (Martin and Kearney 2002) and the B-1 B cell (Hayakawa and Hardy 2000) that are responsible for the responses to TI antigens (see also chapters by K.R. Alugupalli and N. Baumgarth et al., this volume). Recent data demonstrates that marginal zone B cells are also involved in extrafollicular B cell responses to proteins and polysaccharide antigens and are fully capable of initiating germinal center formation and somatic hypermutation (Hayward and Lawton 1977; MacLennan et al. 2003; Sidman and Unanue 1975). The B-1 B cell is thought to be responsible for natural circulating Ig, much of which has anti-bacterial specificity, and the B-1b B cell is responsible for active immunity to polysaccharide antigens (Hayakawa and Hardy 2000).

Fig. 1 (continued) vaccine are then presented to antigen-specific T cells on the cell surface of the B cells in the context of MHC class II. The peptide/MHC complexes are recognized by T cell receptors (TCR) on TH1 and TH2 T cells and T cell activation ensues. The resulting activated T cells provide cognate T cell help to the polysaccharide-specific B cell and the cytokines that are secreted by these activated T cells provide noncognate help to the same PS-specific B cells, which results in the production of PS-specific antibodies. Without the presence of the conjugated protein on the polysaccharide, the B cells specific for the polysaccharide antigens would not receive the ancillary signals from activated T cells and there would be significantly less clonal expansion of the cells, diminished immunological memory, diminished antibody productions, and restricted avidity of the antibody

The role of T cells in the response to intact bacterium has only recently been studied in detail. While early studies have demonstrated a role for noncognate T cell help in the immune response to polysaccharides (Mond et al. 1983; Sverremark and Fernandez 1998), more recently it has been shown that the anti-polysaccharide response to intact bacterium is dependent on α/β CD4$^+$ T cells and more specifically on B7-2/CD28-dependent costimulation as well as on CD40–CD40L interactions (Snapper et al. 2001; Wu et al. 1999, 2000). Despite the apparent classical role of T cells in the anti-polysaccharide response to the organism, there was no induction of immunologic memory as might have been expected. This was shown by Snapper and colleagues to reflect the relatively shorter duration of T cell help for the anti-polysaccharide response as compared to the longer duration of T cell help required for the TD anti-protein response (Snapper et al. 2001). In recent studies using two photon microscopy and intravital staining of explanted lymph nodes, it was shown that there was prolonged contact of B and T cells in the generation of classical T cell-dependent responses (Miller et al. 2002); whether the duration of such contact is shortened for B–T interactions in the response to polysaccharides remains to be studied. It seems clear, however, that the duration of cellular contact and the duration of BCR engagement both have dramatic effects on the ultimate activation of the B cell. While the role of antigen-presenting dendritic cells in the response to intact organism is clear (Colino et al. 2002), the phenotype of the cells and their differential role in the anti-PS vs anti-protein response is currently being studied.

Since all bacterial organisms express PAMPS (pathogen-associated molecular patterns) and thus bind to TLR on phagocytic cells and on B cells, it would be expected that the resultant cytokine secretion would play a role in the subsequent immune response. Pro-inflammatory cytokines could activate dendritic cells at the site of pathogen encounter and mediate cell maturation thereby enhancing activation of CD4$^+$ T cells. In this regard it was shown that TNF-α that is released early after infection with the bacteria was required for induction of optimal anti-protein and anti-polysaccharide responses (Khan et al. 2002). The requirement for TLR signaling, however, may well be specific for different pathogens. Thus in the response to *S. pneumoniae* type 14, MyD88-negative but not TLR2-negative mice were more sensitive to killing following challenge with live Pn14 relative to wild type mice (Chen et al. 2006). In contrast, a recent report demonstrated a role for endogenous TLR2 in stimulating an Ig response to *Haemophilus influenzae* type b outer membrane complex glycoconjugate vaccine (Latz et al. 2004). A role for TLR in response to bacteria is not absolute, however, and a recent publication by Gavin et al. demonstrated that the enhanced antibody responses induced by adjuvants containing bacterial derived components can occur in the absence of TLR signaling (Gavin et al. 2006). It would therefore be expected that bacteria could also stimulate TLR-independent mechanisms leading to antibody responses. Further definition of the role of ancillary signals in the response to TI antigens and to intact bacterium will be important as new adjuvants are being designed to enhance immune responses for the neonatal population and the geriatric population. In this regard, attention needs to be directed at the NK T cell and the γ/δ T cell, both of which recognize polysaccharide or glycolipid antigens (Konigshofer and Chien 2006; Treiner and Lantz 2006) and their recruitment into the response may be important in allowing optimal activation induced by these non-protein antigens. NK T cells can be stimulated in a CD1-dependent manner by the cell wall glycolipid, which are expressed by Gram-negative bacteria (Mattner et al. 2005). They can also be stimulated in a TLR-dependent manner via the upregulation of the endogenous TLR ligand on dendritic cells (Treiner and Lantz 2006). Multiple cytokines which can directly or indirectly affect B cell activation have been shown to be secreted by these cells (Akbari et al. 2003). The other set of nonclassical T cells, the γ/δ set, are also rich sources of B cell stimulatory cytokines and are stimulated by nonclassical antigens. These cells can be stimulated by phosphoantigens and show in vitro responses to phosphoantigens produced by prokaryotes and eukaryotes. Although they constitute only 1%–5% of the T cells in lymphoid organs, they are noted for being overrepresented in the blood of patients during bacterial and viral infections, where they can constitute up to 50% of total T cells (Konigshofer and Chien 2006). This is another source of supplying ancillary help during a CD4 T cell-independent response.

5 Commercial Vaccine Efforts Targeting T Cell-Independent Antigens

Over the last few decades, several vaccines designed to protect neonates and children from infection with encapsulated bacteria have been marketed and all of them target T cell-independent antigens. These vaccines are composed of capsular polysaccharides conjugated to carrier proteins. For induction of protection in the adult population, immunization with the purified polysaccharide is sufficient and conjugate vaccines are not required. Since polysaccharides are antigenically diverse and vary immunologically between different bacterial strains, vaccines may include multiple polysaccharide types to provide the broadest protection from infection. The commercial vaccines that target T cell-independent antigens that will be discussed in this section include vaccines for *Haemophilus influenzae* type b, *Streptococcus pneumoniae*, *Neisseria meningitides*, and *Staphylococcus aureus* (see Tables 1 and 2).

5.1 Haemophilus Influenzae Type B

Haemophilus influenzae type b (Hib) is an encapsulated Gram-negative bacteria responsible for several invasive bacterial diseases in children, including meningitis, septicemia, and pneumonia (Herbert and Moxon 1999). The polyribosyl ribitol phosphate (PRP) capsule of type b *H. influenzae* facilitates survival of the bacteria in the blood through resistance to phagocytosis and complement-mediated killing (Kelly et al. 2004). As early as the 1930s, it was shown that

Table 1 Commercially available vaccines which target T cell-independent antigens

Vaccine	Company	Pathogen	Composition	Web site
Polysaccharide vaccines				
Menomune-A/C/Y/W-135	Sanofi-Pasteur	Neisseria meningitidis	Capsule types A, C, Y and W-135	www.sanofipasteur.com
ACWY Vax	GlaxoSmithKline	Neisseria meningitidis	Capsule types A, C, Y and W-135	www.gsk.com
Pneumovax 23	Merck & Company	Streptococcus pneumoniae	23 pneumococcal polysaccharides	www.merckvaccines.com
Typherix	GlaxoSmithKline	Salmonella typhi	Vi cell surface polysaccharide	www.gsk.com
Typhim Vi	Sanofi-Pasteur	Salmonella typhi	Surface Vi polysaccharide	www.sanofipasteur.com
Conjugate vaccines				
ActHIB	Sanofi-Pasteur	Haemophilus influenzae type B	PRP polysaccharide/ tetanus toxoid	www.sanofipasteur.com
Hiberix	GlaxoSmithKline	Haemophilus influenzae type B	PRP polysaccharide/ tetanus toxoid	www.gsk.com
HibTITER	Wyeth	Haemophilus influenzae type B	PRP oligosaccharide bound to CRM197	www.wyeth.com
PedvaxHIB	Merck & Company	Haemophilus influenzae type B	PRP/outer membrane complex from N. meningitides type B	www.merckvaccines.com
Menjugate	Novartis AG	Neisseria meningitides type C	O-acetylated polysaccharide/ CRM197	www.novartis.com
Meningitec	Wyeth	Neisseria meningitides type C	O-acetylated polysaccharide/ CRM197	www.wyeth.com
NeisVac-C	Baxter Vaccines	Neisseria meningitides type C	deacylated polysaccharide/ tetanus toxoid	www.baxtervaccines.com
Menactra	Sanofi-Pasteur	N. meningitides types A, C, Y and W-135	Polysaccharide/ diphtheria toxoid	www.sanofipasteur.com
Prevnar	Wyeth	Streptococcus pneumoniae	Seven polysaccharides/ diphtheria toxoid	www.wyeth.com

anti-polysaccharide antibodies were protective against invasive disease (Kelly et al. 2004). Unfortunately, the PRP capsule by itself is poorly immunogenic, particularly in infants and young children (a population at high risk for invasive disease) because of their immunologic immaturity, and fails to induce either a primary response or immunologic memory (Kelly et al. 2004). Conjugation of PRP to various protein carriers, however, overcame this unresponsiveness to the

Table 2 Late-stage developmental vaccines and antibody therapies which target T cell independent antigens

Vaccine	Company	Pathogen	Composition	Web site
Conjugate vaccines				
Streptorix	GlaxoSmithKline	*Streptococcus pneumoniae*	Multi-polysaccharide conjugate	www.gsk.com
13vPnC	Wyeth	*Streptococcus pneumoniae*	13 polysaccharide conjugate vaccine	www.wyeth.com
Menitorix	GlaxoSmithKline	*Neisseria meningitis* group C disease & *Haemophilus influenzae* type b disease prophylaxis	Multi-component vaccine	www.gsk.com
Globorix	GlaxoSmithKline	Diphtheria, tetanus, pertussis, hepatitis B, *Haemophilus influenzae* type b disease, *Neisseria meningitis* groups	Multi-component vaccine	www.gsk.com
MenACWY	Novartis AG	*Neisseria meningitis* groups A,C,W and Y	Polysaccharides from types A, C, Y and W-135 conjugate	www.novartis.com
StaphVAX	NABI Biopharmaceuticals	*Staphylococcus aureus*	Types 5 and 8 polysaccharide /*P. aeruginosa* exotoxin A	www.nabi.com
Antibody therapies				
AltaStaph	NABI Biopharmaceuticals	*Staphylococcus aureus*	Polyclonal antibody preparation from StaphVAX immunized volunteers	www.nabi.com
BSYX-A110 (Pagibaximab)	Biosynexus Incorporated	*Staphylococci*	Humanized monoclonal antibody to lipoteichoic acid	www.biosynexus.com

vaccine and was effective in inducing immunity and immunologic memory in children (Kelly et al. 2004). Covalent linkage of the polysaccharide to the protein was shown by many groups to be essential for inducing enhanced immunity to the polysaccharide antigen in animal models, and this provided the basis for designing protein–polysaccharide conjugate vaccines for the neonatal population (Kelly et al. 2004). In a more recent publication (Latz et al. 2004), it was demonstrated that unlike PRP conjugated to the protein carriers tetanus toxoid or CRM197 protein, PRP conjugated to the outer membrane protein complex of *N. meningitides*, which enables binding to TLR2 on dendritic cells, appears to be more immunogenic in infants after a single dose.

The first commercially available conjugate vaccines were vaccines for protection from *H. influenzae* type b, and these vaccines were introduced into North America, Finland, and the United Kingdom in the late 1980s and early 1990s (Kelly et al. 2004). There are six capsular types of *H. influenzae*, but type b causes the majority of invasive disease. The type b capsular polysaccharide PRP is covalently conjugated to various protein carriers to form the vaccines. Several commercial conjugate vaccines are now available and these include ActHIB (Sanofi-Pasteur 2005a), a vaccine from Pasteur Merieux (now Sonofi Pasteur), which is native high-molecular-weight PRP polysaccharide conjugated to tetanus toxoid; GlaxoSmithKline produces Hiberix (GlaxoSmithKline 2004), which is also polysaccharide conjugated to tetanus toxoid; HibTITER (Wyeth 2005) from Lederle (now Wyeth) is PRP oligosaccharide bound to CRM197, which is a mutant diphtheria toxoid; and PedvaxHIB (Merck 2006) from Merck is native PRP attached to the outer membrane complex of serogroup b *N. meningitides*. Merck also produces COMVAX, which is a combination vaccine against Hib and hepatitis B virus, while Sanofi-Pasteur also produces TriHIBit, a Haemophilus b Conjugate Vaccine (Tetanus Toxoid Conjugate)-ActHIB Reconstituted with Diphtheria and Tetanus Toxoids and Acellular Pertussis Vaccine Adsorbed-Tripedia. All of these vaccines are designed to be administered to infants for the prevention of invasive disease and have demonstrated impressive protective efficacy (Kelly et al. 2004). The conjugate vaccines are administered beginning the first few months of life as a course of two or three intramuscular injections. In North America, a further booster is recommended at 12–15 months of age (Kelly et al. 2004). The use of the Hib conjugate vaccines over the past 15 years have virtually eliminated invasive Hib disease in the industrialized world (Kelly et al. 2004).

While Hib conjugate vaccines have proven to be very effective since their introduction for protection from invasive bacterial infection, there are some limitations to these vaccines. Since 1999, there have been some true vaccine failures documented for Hib polysaccharide conjugate vaccines in the UK (McVernon et al. 2004). These failures may be linked to the increased use of acellular pertussis combination vaccines starting in 1999 (Kelly et al. 2004). There appears to be a difference in induction of memory responses among vaccinated children, and children with invasive Hib disease, whether vaccinated or not, have immunologic responses consistent with immunologic memory (Kelly et al. 2004). This finding demonstrates that immunologic memory may not necessarily be an indicator of protection. If serum antibody levels fall below a protective threshold, it may take several days for antibody levels to reach protective levels as immunologic memory comes into play upon challenge with an invading pathogen. This delay may allow invasive disease to occur despite proper vaccination. Correlates between measures of immunologic memory and protection have not yet been defined. These findings have prompted a recommendation for an additional Hib booster to be administered between 6 months and 4 years of age (Kelly et al. 2004). This waning immunity may also come into play later in life when older people begin to be at higher risk for invasive bacterial infections.

Vaccine development to protect against nontypeable *H. influenzae* is ongoing. Nontypeable *H. influenzae* is responsible for a large proportion of acute otitis media and chronic sinusitis throughout the world (Barenkamp 2004), and vaccine efforts have largely focused upon identifying the appropriate protein surface antigens as nontypeable *H. influenzae* lacks a polysaccharide capsule. In addition to proteins, candidate antigens also include T cell-independent antigens such as detoxified lipid A coupled to a protein carrier or core oligosaccharide common to lipooligosaccharide of all nontypeable *H. influenzae* (Barenkamp 2004). Reports indicate that conjugation of lipoprotein D to polysaccharide confers significant protection to infection with nontypeable *H. influenzae* (Poolman et al. 2000).

5.2 *Neisseria Meningitides*

Neisseria meningitides is another encapsulated Gram-negative bacteria (Cartwright 1999) for which vaccines targeting T cell-independent antigens have proven protective (Girard et al. 2006). There are 13 distinct serogroups of meningococci based on their polysaccharide capsules, but groups A, B, C, Y, and W-135 are responsible for over 90% of what is considered to be of the most serious infections in young people, meningitis (Girard et al. 2006; Ruggeberg and Pollard 2004). The various serotypes of meningococci are due to the high degree of antigenic variability of the capsular polysaccharides, and polysaccharide types demonstrate some geographic specificity (Girard et al. 2006; Ruggeberg and Pollard 2004). Group A meningococci have been responsible for large-scale epidemics of meningitis in developing countries, but rarely cause disease in the developed world (Girard et al. 2006). Type B meningococci are the most important cause of endemic meningitis in the industrialized world, with types C and Y being the next largest contributors in the US (Danzig 2004; Girard et al. 2006). As early as 1969, the protective effect of bactericidal antibodies against the type C capsular polysaccharide of *N. meningitides* was demonstrated in military recruits (Girard et al. 2006). The first meningococcal vaccine was anti-types A and C vaccine consisting of purified polysaccharides, which was developed to protect military recruits (Girard et al. 2006; Ruggeberg and Pollard 2004). A third and fourth serogroup, W-135 and Y, have also been added to the original bivalent vaccine to create a tetravalent purified polysaccharide vaccine, which is available worldwide. Commercial purified polysaccharide vaccines are available from Sanofi-Pasteur (Menomune-A/C/Y/W-135) and GSK (ACWY Vax; GlaxoSmithKline 2002a).These two purified polysaccharide vaccines are not particularly effective in children and fail to induce a memory response due to their T cell-independent nature (Danzig 2004; Girard et al. 2006). Despite these shortcomings, these polysaccharide vaccines have proven to be effective for a number of indications, including protection from close contact during sporadic cases, mass immunization of closed communities, mass immunization during prolonged outbreaks, and protection of travelers to endemic areas (Girard et al. 2006). Further, the Advisory Committee on Immunization Practices (ACIP) recommended in 2000 that all entering college students be informed of the risk of meningitis and the availability of the polysaccharide vaccine (Girard et al. 2006). Routine vaccination with meningococcal polysaccharide vaccines, however, are not recommended by the ACIP (Danzig 2004).

As was the case with Hib conjugate vaccines, greater success of immunization against *N. meningitides* was found when using polysaccharide protein conjugate vaccines. Effective glycoconjugate vaccines for meningococci are best prepared from low-molecular-weight oligosaccharides (Girard et al. 2006). Meningococcal group C conjugate vaccines have been developed by Chiron (now part of Novartis) (Menjugate; (Region-of-Waterloo-Public-Health 2004) and Wyeth (Meningitec; Wyeth-Australia 2005), both of which are type C polysaccharide conjugated to CRM197, and Baxter (NeisVac-C; Baxter-Healthcare-Australia 2003), which is type C polysaccharide conjugated to tetanus toxoid. The Chiron and Wyeth vaccines have O-acetylated polysaccharides, while the Baxter vaccine is deacetylated. Three doses of vaccine are administered at monthly intervals from 2 months of age, with some studies including a booster at a later point (Ruggeberg and Pollard 2004). The type C vaccine was introduced first in Europe and Sanofi-Pasteur now has a quadravalent meningococcal conjugate vaccine available in the US as Menactra (Sanofi-Pasteur 2006), which consists of meningococcal groups A, C, Y, and W-135 polysaccharides conjugated to diphtheria toxoid. Within 1 year of introduction of the type C glycoconjugate vaccine in the UK, an 85% reduction in the number of confirmed cases of type C meningococcal disease was seen (Girard et al. 2006).

Group B meningococci cause a substantial portion of meningococcal infections throughout the world, but this group of meningococci can not be prevented by capsular polysaccharide-based vaccines (Danzig 2004; Girard et al. 2006; Ruggeberg and Pollard 2004). The correlation between serum bactericidal activity and anti-polysaccharide IgG levels is poor, especially in children (Girard et al. 2006). Further, the group B polysaccharide, $\alpha(2-8)$-linked N-acetyl-neuraminic acid (polysyalic acid), is also present in human tissue, including neural cells, leading to safety concerns about an antibody response which could potentially cross-react with human tissue (Girard et al. 2006). Sanofi-Pasteur tested a vaccine candidate which replaced the N-acetyl groups of the group B polysaccharide with N-propinyl groups, and while the vaccine was found to be safe, the antibodies lacked functional activity. Group B vaccine research has now turned to cell surface proteins contained in outer-membrane vesicles as vaccine targets.

5.3 *Streptococcus Pneumoniae*

Streptococcus pneumoniae is an encapsulated Gram-positive organism that is a major cause of invasive diseases such as meningitis, septicemia, and pneumonia. *S. pneumoniae* is also a main cause of acute otitis media and sinusitis (Briles et al. 2000). Colonization of the nasopharyngeal tissue is common, especially in circumstances of crowding such as day-care centers. This colonization does not necessarily lead to disease, but perturbances of the immune system can lead to localized infection (e.g., an ear infection as a sequela to a cold) or invasive disease when there is damage to the local mucosa. The resistance to penicillin is so widespread in *S. pneumoniae* that there were early and intensive efforts to develop an effective vaccine. Early vaccine efforts concentrated on purified capsular polysaccharides, but pneumococcal capsule can consist of many different polysaccharide types (Briles et al. 2000). As early as 1911, whole killed pneumococci were tested as vaccine candidates, but these efforts only raised an antibody response against one of two prevalent capsular types and failed to provide broad protection (Bogaert et al. 2004). Isolated polysaccharides from *S. pneumoniae* were first successfully tested as vaccines in 1926 (Bogaert et al. 2004). There are many antigenically divergent capsular polysaccharide types found in pneumococci, and successful vaccines must include many of these polysaccharide types (Briles et al. 2000). A 14-valent purified polysaccharide vaccine was first produced in 1977 and was later expanded to 23-valent in 1983 (Bogaert et al. 2004). As was found with other purified polysaccharide vaccines, this vaccine was not particularly immunogenic in small children or immunocompromised people. Currently, there is only one commercially available vaccine containing purified pneumococcal polysaccharides from a large vaccine production company, and this vaccine is Pneumovax 23 (Merck 2005) produced by Merck & Company. The vaccine consists of 23 purified pneumococcal polysaccharides which account for more than 85% of pneumococcal infections (Bogaert et al. 2004). This vaccine is effective in adults and children older than 2 years, and may also be effective against acute pneumococcal otitis media (Bogaert et al. 2004). The Advisory Committee on Immunization Practices recommends this vaccine only for certain groups that are at increased risk for pneumococcal disease (Bogaert et al. 2004). One group at increased risk for pneumococcal infection is HIV-infected individuals, but these individuals mount a considerably lower peak IgG response after vaccination. This reduced peak is not consistent across all pneumococcal serotypes included in the vaccine, however, suggesting that some pneumococcal polysaccharide types may have a T cell-dependent immune contribution which is impaired in the HIV-positive patient population (Bogaert et al. 2004).

Conjugation of pneumococcal polysaccharides to carrier proteins has also been an effective strategy for generating a T cell-dependent response to these T cell-independent antigens and has lead to improved primary immune responses and increased memory. Pneumococci are unique in the vast number of polysaccharide variations expressed in their capsules (Briles et al. 2000). The number of polysaccharide types which can be conjugated to a single carrier protein or administered as a single formulation is limited, however. Thus far, only Prevnar from Wyeth (Wyeth 2006) has been approved for use as a conjugate vaccine for protection from invasive pneumococcal disease. This conjugate vaccine is septavalent and has coverage for between 55% and 80% of pneumococcal isolates for invasive disease depending on geographic local (Bogaert et al. 2004). This vaccine may also lead to a reduction in nasopharyngeal colonization, possibly because the conjugate vaccines generate both systemic and mucosal antibodies (Bogaert et al. 2004). The ACIP recommends the pneumococcal conjugate vaccine for all children under 24 months. Unfortunately, as with the purified polysaccharide vaccine, the conjugate vaccine was also found to have limited effectiveness against acute otitis media. In a 2005 report titled "Impact of Conjugate Pneumococcal Vaccines" (Whitney 2005), it was determined that the "Pneumococcal conjugate vaccines have been highly successful in reducing the incidence of invasive pneumococcal disease in clinical trails." GSK is also developing Streptorix, a pneumococcal conjugate vaccine which incorporates more polysaccharide types and a protein carrier which should also provide protection against nontypeable *H. influenzae*. This vaccine is currently in phase III clinical trials.

5.4 Staphylococcus Aureus

S. aureus is one of the leading causes of nosocomial infection and has a high mortality rate. The search for an effective *S. aureus* vaccine has further been fueled by emergence of multidrug resistance strains of *S. aureus*, including community-associated strains of methicillin-resistant *S. aureus* and fully vancomycin-resistant strains of *S. aureus* (Lowy 2003). One recent disappointment in the area of glycoconjugate vaccines is the recent failure of the clinical trials designed to test the efficacy of capsular polysaccharide conjugate vaccines for protection from *Staphylococcus aureus*. This vaccine, StaphVAX (Fattom et al. 2004), was under development by NABI Biopharmaceuticals (www.nabi.com) and consists of purified capsular polysaccharide of types 5 and 8 (the two most prevalent *S. aureus* capsular types; Lee and Lee 2000) conjugated to *Pseudomonas aeruginosa* exotoxin A (Fattom et al. 2004). The search for an effective vaccine against *S. aureus* has run the gamut from using whole killed bacteria to using purified virulence factors such as adhesins and toxins (Lee 1996).

It was recognized early in staphylococcal research that opsonic antibodies to *S. aureus* were important for protection from infection (Fattom et al. 2004). Many antigen targets have been investigated for their capacity to raise such antibodies (Lee 1996). Development of *S. aureus* typing serum in the 1980s identified as many as 13 capsule types for *S. aureus*, but *S. aureus* expressing two capsule types, 5 and 8, account for as much as 93% of disease isolates from different countries (Fattom et al. 2004; Lee and Lee 2000; Riordan and Lee 2004). Purified polysaccharide from capsule types 5 and 8 are poorly immunogenic in mice by themselves, but coupling these polysaccharides to carrier proteins increases their immunogenicity (Fattom et al. 2004). Further, the conjugate vaccine had T cell-dependent properties, as demonstrated by a booster response after second immunizations. Antibodies generated by this conjugate vaccine were also opsonic in vitro and protective in animal models (Fattom et al. 2004).

StaphVAX, the commercial version of the staphylococcal conjugate vaccine, was tested in various clinical trails (Fattom et al. 2004). In phase 1 trials, the individual conjugate vaccines (capsule type 5 or type 8) were found to be safe and immunogenic in healthy adult volunteers. Most volunteers had a background antibody titer to *S. aureus* capsular polysaccharide, probably due to the fact that many people are either permanently or transiently nasally colonized with *S. aureus* during their life, and this background titer was boosted substantially by the first dose of vaccine. A second dose of vaccine given 6 weeks after the first dose did not lead to further increase in antibody titer, however. Sera from immunized volunteers was found to have type-specific opsonophagocytic activity. In subsequent clinical trails when types 5 and 8 capsular polysaccharide conjugates were combined in one injection, similar immune responses were seen in response to the combination as were seen to the individual components (Fattom et al. 2004). A phase III efficacy trial was conducted and 1800 end-stage renal disease patients on hemodialysis received either vaccine or placebo control (Fattom et al. 2004). The primary endpoint for the trial was a significant reduction of bacteremia for 1 year, and while a single immunization induced a good antibody response to both capsule types, the trial failed to meet the primary endpoint to statistical significance at 1 year. Post-hoc analysis did demonstrate a statistically significant reduction in bacteremia at 32 and 40 weeks after immunization, however. A confirmatory phase III clinical trial was conducted in which a booster immunization was administered at 8 months and the extension of efficacy was evaluated. As described on the company web site (www.nabi.com), "In November 2005, the Company announced that the StaphVAX Phase III confirmatory clinical trial failed to meet its primary endpoint. The study found no reduction in *S. aureus* types 5 and 8 infections in the StaphVAX group as compared to the placebo group. The study was a randomized, double-blinded, placebo-controlled trial of 3,600 patients on hemodialysis. The Company is assessing factors that may have contributed to this result and the process will take several months. As a result, the Company has placed further clinical trial development of StaphVAX and Altastaph [*Staphylococcus aureus* Immune Globulin Intravenous (Human)] on hold until the assessment is completed. The Company has also withdrawn its Marketing Authorization Application (MAA) to market StaphVAX in the European Union." Further assessment by Nabi of factors that contributed to the phase III clinical failure concluded that, "The assessment revealed marked differences in the effectiveness of the lot of vaccine used in this [the confirmatory] study compared to the lot used in the previous Phase III clinical study". Further, "the quality or functional characteristics of the antibodies generated by the vaccine used in the confirmatory clinical study were inferior to those antibodies generated by vaccine lots used in previous and subsequent clinical studies." Also, "medical factors associated with kidney disease in dialysis patients impaired their immune response to the vaccine. When considered in combination with an increase in the virulence of the bacteria, these factors also contributed to the observed lack of protection in this study population." According to the company web site (accessed January 2007), the company plans to continue to develop and test StaphVAX, adding additional polysaccharide immunogens for *S. aureus* as well as *S. epidermidis*, and seek a strategic partner for this endeavor. They also intend to accelerate the development of Altastaph, which is a polyclonal antibody preparation purified from volunteers immunized with StaphVAX.

While previous success with conjugate vaccines for other pathogens might suggest that this strategy should have a good chance of success, *S. aureus* has some significant differences from these other pathogens. For example, while polysaccharide capsule is a primary virulence factor of the other pathogens, protecting them from opsonophagocytosis and complement-mediated killing, there has been some question as to whether expression of capsule truly contributes to *S. aureus* pathogenesis in all infections. *S. aureus* mutants unable to express capsule are still pathogenic in some models (Riordan and Lee 2004), and whether *S. aureus* actually expresses capsule in all *S. aureus* infections in humans remains unclear. Further, *S. aureus* is a prodigious toxin producer, expressing numerous factors that can attenuate an immune response (Bohach and Foster 2000). Finally, staphylococci also form biofilms on both indwelling devices and damaged tissue (Donlan and Costerton 2002). Once the staphylococci have established a biofilm, it would be difficult for an immune response to a vaccine or passively administered antibodies to clear this biofilm. It may be important for any vaccine strategy for staphylococci to ensure that protective antibody titers are established prior to the implantation of an indwelling device, such as a dialysis port where the staphylococci can establish a biofilm.

Antibody-based therapy targeting other staphylococcal T cell-independent antigens are also under development. BSYX-A110 (pagibaximab, Biosynexus Incorporated, www.biosynexus.com) is a chimeric monoclonal antibody against lipoteichoic acid (LTA) of staphylococci. LTA is an essential component of the staphylococcal cell wall (Neuhaus and Baddiley 2003); and LTA-negative staphylococcal mutants are believed to be nonviable and have never been isolated (Fischer 1994). BSYX-A110 is currently under development for the prevention of staphylococcal sepsis in premature low-birth-weight infants. Results of a number of clinical trials run in premature low-birth-weight infants are very encouraging both in terms of the safety profile of the antibody as well as in its ability to achieve high levels of opsonic activity in the infused babies. Monoclonal antibodies to LTA were generated by immunizing mice with whole killed *S. epidermidis* (strain Hay) in which the T cell-independent LTA antigen was presented to the murine immune system in the context of whole bacteria, similar in some respects to conjugation to a carrier protein. It may also be possible to use LTA as an antigen in a staphylococcal vaccine, perhaps presenting the molecule as a protein conjugate, and this possibility is being investigated. Wall teichoic acid (WTA), another cell surface glycolipid may also serve as an attractive T cell-independent target for anti-*S. aureus* antibody therapy or vaccines. WTA has been shown to be essential for *S. aureus* nasal colonization in animal models and may actually be involved in the initial binding of the *S. aureus* to nasal tissue (Weidenmaier et al. 2004). Antibodies against this T cell-independent antigen may serve to block the ligand–receptor interaction and thus block nasal colonization. Other T cell-independent antigens, beyond capsular polysaccharides, may also prove to be attractive targets for vaccine antigens or antibody-mediated therapies in other bacteria.

5.5 *Salmonella Typhi*

A less commonly used type of commercial vaccine which contains T cell-independent antigens are vaccines that target typhoid fever. GSK produces Typherix (GlaxoSmithKline 2002b), which contains the Vi cell surface polysaccharide extracted from the *Salmonella typhi* Ty 2 strain. This is an unconjugated polysaccharide vaccine developed for travelers to areas endemic for typhoid fever. Sanofi-Pasteur also produces Typhim Vi (Sanofi-Pasteur 2005b), which is cell surface Vi polysaccharide extracted from the *Salmonella enterica* serovar Typhi, S typhi Ty2 strain. This purified polysaccharide vaccine is also for travelers to endemic areas and is expected to provide protection for up to 3 years.

6 Conclusion

The introduction of conjugate vaccines has eliminated many bacterial infections that infect the pediatric age population. Fortunately, antibody pressure has not selected for bacteria that have mutated to elude antibody-mediated opsonization, thus rendering the vaccines obsolete. The absence of mutation in response to this type of pressure likely results from the fact that the capsule of these pathogens is both necessary for survival of the organism and is composed of polysaccharides produced by the protein products of multigene operons which would be very difficult to change by DNA mutation. It remains to be seen whether antibody pressure on certain strains of bacteria such as S. *pneumonia* might select for the emergence of additional strains which heretofore were not carried in the nasopharynx; this could conceivably alter the identity of potentially infecting strains of bacteria. Unfortunately, conjugate vaccines have not proven to be successful in protecting the adult and the elderly population. The challenge of the future lies in designing conjugate vaccines that will be effective in the elderly population, which is also an immune-suppressed group of individuals. The introduction of new and safe adjuvants into the vaccine field might enable vaccine manufacturers to combine many vaccine targets in one vial without the concern for antigenic competition. This would minimize the number of injections that need to be given to children. Additionally, adjuvants may allow for fewer injections to be given because of the enhanced immunity induced in the neonatal population. These two last points would be extremely important advances for the developing world where medical visits are few and medical costs are high and thus reduced inoculations could increase the compliance and protection in the population requiring these conjugate vaccines.

References

Ahmad A, Mond JJ (1986) Restoration of in vitro responsiveness of *xid* B cells to TNP-Ficoll by 8-mercaptoguanosine. J Immunol 136:1223–1226

Akbari O, Stock P, Meyer E, Kronenberg M, Sidobre S, Nakayama T, Taniguchi M, Grusby MJ, DeKruyff RH, Umetsu DT (2003) Essential role of NKT cells producing IL-4 and IL-13 in the development of allergen-induced airway hyperactivity. Nat Med 9:582–588

AlonsoDeVelasco E, Verheul AF, Verhoef J, Snippe H (1995) *Streptococcus pneumoniae*: virulence factors, pathogenesis, and vaccines. Microbiol Rev 59:591–603

Balazs M, Martin F, Zhou T, Kearney JF (2002) Blood dendritic cells interact with splenic marginal zone B cells to initiate T-independent immune responses. Immunity 17:341–352

Barenkamp SJ (2004) Rationale and prospects for a nontypable *Haemophilus influenzae* vaccine. Pediatr Infect Dis J 23:461–462

Baril L, Dietemann J, Essevaz-Roulet M, Beniguel L, Coan P, Briles DE, Guy B, Cozon G (2006) Pneumococcal surface protein A (PspA) is effective at eliciting T cell-mediated responses during invasive pneumococcal disease in adults. Clin Exp Immunol 145:277–286

Baxter-Healthcare-Australia (2003) NeisVac-C™ Vaccine. Consumer Medicine Information, NeisVac-C

Blery M, Tze L, Miosge LA, Jun JE, Goodnow CC (2006) Essential role of membrane cholesterol in accelerated BCR internalization and uncoupling from NF-kappa B in B cell clonal anergy. J Exp Med 203:1773–1783

Bogaert D, Hermans PWM, Padrian PV, Rumke HC, de Groot R (2004) Pneumococcal vaccines: an update on current strategies. Vaccine 22:2209–2220

Bohach GA, Foster TJ (2000) *Staphylococcus aureus* exotoxins. In: Fischetti VA, Novick RP, Ferretti JJ, Portnoy DA, Rood JI (eds) Gram-positive pathogens. ASM Press, Washington DC, pp 367–378

Boswell CM, Stein KE (1996) Avidity maturation, repertoire shift, and strain differences in antibodies to bacterial levan, a type 2 thymus-independent polysaccharide antigen. J Immunol 157:1996–2005

Briles DE, Patton JC, Swialto E, Nahm MH (2000) Pneumococcal vaccines. In: Fischetti VA, Novick RP, Ferretti JJ, Portnoy DA, Rood JI (eds) Gram-positive pathogens. ASM Press, Washington DC, pp 244–250

Brunswick M, Finkelman FD, Highet PF, Inman JK, Dintzis HM, Mond JJ (1988) Picogram quantities of anti-Ig antibodies coupled to dextran induce B cell proliferation. J Immunol 140:3364–3372

Brunswick M, June CH, Finkelman FD, Dintzis HM, Inman JK, Mond JJ (1989) Surface Ig-mediated B cell activation in the absence of detectable elevations in intracellular ionized calcium: a model for T-cell-independent B-cell activation. Proc Natl Acad Sci U S A 86:6724–6728

Cartwright KA (1999) *Neisseria meningitidis*. In: Yu VL, Merigan TC, Barriere SL (eds) Antimicrobial Therapy and Vaccines. Williams & Wilkins, Baltimore, pp 303–309

Casal J, Tarrago D (2003) Immunity to *Streptococcus pneumoniae*: factors affecting production and efficacy. Curr Opin Infect Dis 16:219–224

Chen Q, Sen G, Snapper CM (2006) Endogenous IL-1R1 signaling is critical for cognate CD4+ cell help for introduction of in vivo type 1 and type 2 antipolysaccharide and antiprotein Ig isotype responses to intact *Streptococcus pneumoniae* but not to a soluble pneumococcal conjugate vaccine. J Immunol 177:6044–6051

Colino J, Shen Y, Snapper CM (2002) Dendritic cells pulsed with intact *Streptococcus pneumoniae* elicit both protein- and polysaccharide-specific immunoglobulin isotype responses in vivo through distinct mechanisms. J Exp Med 195:1–13

Danzig L (2004) Meningococcal vaccines. Pediatr Infect Dis J 23:S285–S292

Dempsey PW, Allison ME, Akaraju S, Goodenow C, Fearon DT (1996) C3d of complement as a molecular adjuvant: bridging innate and acquired immunity. Science 271:348–350

Donlan RM, Costerton JW (2002) Biofilms: survival mechanisms of clinically relevant microorganisms. Clin Microbiol Rev 15:167–193

Donnelly JJ, Deck RR, Liu MA (1990) Immunogenicity of a *Haemophilus influenzae* polysaccharide-*Neisseria meningitidis* outer membrane protein complex conjugate vaccine. J Immunol 145:3071–3079

Fattom AI, Horwith G, Fuller S, Propst M, Naso R (2004) Development of StaphVAX®, a polysaccharide conjugate vaccine against *S. aureus* infection: from lab bench to phase III clinical trials. Vaccine 22:880–887

Fischer W (1994) Lipoteichoic acid and lipids in the membrane of *Staphylococcus aureus*. Med Microbiol Immunol 183:61–76

Fuentes-Panana EM, Bannish G, Monroe JG (2004) Basal B-cell receptor signaling in B lymphocytes: mechanisms of regulation and role in positive selection, differentiation, and peripheral survival. Immunol Rev 197:26–40

Gavin AL, Hoebe K, Duong B, Ota T, Martin C, Beutler B, Nemazee D (2006) Adjuvant-enhanced antibody responses in the absence of toll-like receptor signaling. Science 314:1936–1938

Girard MP, Preziosi MP, Aguado MT, Kieny MP (2006) A review of vaccine research and development: Meningococcal disease. Vaccine 24:4692–4700

GlaxoSmithKline (2002a) ACWY Vax® Meningococcal Polysaccharide Vaccine PhEur. Patient information leaflet, 45393

GlaxoSmithKline (2002b) Typherix VI® polysaccharide typhoid vaccine. Patient information leaflet, 3153

GlaxoSmithKline (2004) Hiberix® *Haemophilus influenza* type b (Hib) Vaccine. Patient information leaflet, 407575

Golding H, Foiles PG, Rittenberg MB (1982) Partial reconstitution of TNP-Ficoll responses and IgG3 expression in Xid mice undergoing graft-vs-host reaction. J Immunol 129:2641–2646

Hayakawa K, Hardy RR (2000) Development and function of B-1 Cells. Curr Opin Immunol 12:346–353

Hayward AR, Lawton AR (1977) Induction of plasma cell differentiation of human fetal lymphocytes: evidence for functional immaturity of T and B cells. J Immunol 119:1213–1217

Herbert MA, Moxon ER (1999) *Haemophilus influenzae*. In: Yu VL, Merigan TC, Barriere SL (eds) Antimicrobial therapy and vaccines. Williams & Wilkins, Baltimore. pp 213–227

Karnell FG, Brezski RJ, King LB, Silverman MA, Monroe JG (2005) Membrane cholesterol content accounts for developmental differences in surface B cell receptor compartmentalization and signaling. J Biol Chem 280:25621–25628

Kelly DF, Moxon ER, Pollard AJ (2004) *Haemophilus influenzae* type b conjugate vaccines. Immunol Rev 113:163–174

Khan AQ, Shen Y, Wu ZQ, Wynn TA, Snapper CM (2002) Endogenous pro- and anti-inflammatory cytokines differentially regulate an in vivo humoral response to *Streptococcus pneumoniae*. Infect Immun 70:749–761

Khan AQ, Chen Q, Wu ZQ, Paton JC, Snapper CM (2005) Both innate immunity and type 1 humoral immunity to *Streptococcus pneumoniae* are mediated by MyD88 but differ in their relative levels of dependence on toll-like receptor 2. Infect Immun 73:298–307

Khan WN, Alt FW, Gerstein RM, Malynn BA, Larsson I, Rathbun G, Davidson L, Muller S, Kantor AB, Herzenberg LA, Rosen FS, Sideras P (1995) Defective B cell development and function in Btk-deficient mice. Immunity 3:283–299

Koedel U, Angele B, Rupprecht T, Wagner H, Roggenkamp A, Pfister H-W, Kirschning CJ (2003) Toll-like receptor 2 participates in mediation of immune response in experimental pneumococcal meningitis. J Infect Dis 186:798–806

Konigshofer Y, Chien Y (2006) Gamma delta T cells—innate immune lymphocytes? Curr Opin Immunol 18:527–533

Kuo J, Douglas M, Ree HK, Lindberg AA (1995) Characterization of a recombinant pneumolysin and its use as a protein carrier for pneumococcal type 18C conjugate vaccines. Infect Immun 63:2706–2713

Landers CD, Bondada S (2005) CpG oligodeoxynucleotides stimulate cord blood mononuclear cells to produce immunoglobulins. Clin Immunol 116:236–245

Latz E, Franko J, Golenbock DT, Schreiber JR (2004) *Haemophilus influenzae* type b-outer membrane protein complex glyco-conjugate vaccine induces cytokine production by engaging human toll-like receptor (TLR2) and requires TLR2 for optimal immunogenicity. J Immunol 172:2431–2438

Lee CY, Lee JC (2000) Staphylococcal capsule. In: Fischetti VA, Novick RP, Ferretti JJ, Portnoy DA, Rood JI (eds) Gram-positive pathogens. ASM Press, Washington DC, pp 361–366

Lee JC (1996) The prospects for developing a vaccine against *Staphylococcus aureus*. Trends Microbiol 4:162–166

Lowy FD (2003) Antimicrobial resistance: the example of *Staphylococcus aureus*. J Clin Invest 111:1265–1273

MacLennan JC, Toellner KM, Cunningham AF, Serre K, Sze DM (2003) Extra follicular antibody responses. Immunol Rev 194:8–18

Malissein E, Verdier M, Ratinaud MH, Troutaud D (2006) Activation of Bad trafficking is involved in the BCR-mediated apoptosis of immature B cells. Apoptosis 11:1003–1012

Malley R, Henneke P, Morse SC, Cieslewicz MJ, Lipsitch M, Thompson CM, Kurt-Jones E, Paton JC, Wessels MR, Goldenbock DT (2003) Recognition of pneumolysin by Toll-like receptor 4 confers resistance to pneumococcal infection. Proc Natl Acad Sci U S A 100:1966–1971

Martin F, Kearney JF (2002) Marginal zone B cells. Nat Rev Immunol 2:323–335

Mattner J, Bebord KL, Ismail N, Goff RD, Cantu C, Zhou D, Saint-Mezard P, Wang V, Gao Y, Yin N, Hoebe K, Schneewind O, Walker D, Beutler B, Teyton L, Savage PB, Bendelac A (2005) Exogenous and endogenous glycolipid antigens activate NKT cells during microbial infections. Nature 434:525–529

Mawas F, Dickinson R, Douglas-Bardsley A, Xing DK, Sesardic D, Corbel MJ (2006) Immune interaction between components of acellular pertussis-diphtheria-tetanus (DTaP) and *Haemophilus influenzae* b (Hib) conjugate vaccine in a rat model. Vaccine 24:3505–3512

McVernon J, Mitchison NA, Moxon ER (2004) T helper cells and efficacy of *Haemophilus influenzae* type b conjugate vaccination. Lancet Infect Dis 4:40–43

Merck (2005) Pneumovax 23® (Pneumococcal vaccine polyvalent). Product circular, 7999825

Merck (2006) PedvaxHIB® Haemophilus b conjugate vaccine (www.merckvaccines.com/vaccines/haem/index.html)

Miller MJ, Wei SH, Parker I, Cahalan MD (2002) Two-photon imaging of lymphocyte motility and antigen response in intact lymph node. Science 296:1869–1873

Mitchison NA (2004) T-cell-B-cell cooperation. Nat Rev Immunol 4:308–312

Mond JJ, Sehgal E, Sachs DH, Paul WE (1979a) Expression of Ia antigen on adult and neonatal B lymphocytes responsive to thymus-independent antigens. J Immunol 123:1619–1623

Mond JJ, Stein KE, Subbarao B, Paul WE (1979b) Analysis of B cell activation requirements with TNP-conjugated polyacrylamide beads. J Immunol 123:239–245

Mond JJ, Farrar J, Paul WE, Fuller-Farrar J, Schaeffer M, Howard M (1983) T cell dependence and factor reconstitution of In vitro antibody responses to TNP-*B. abortus* and TNP-ficoll: restoration of depleted responses with chromatographed fractions of a T cell-derived factor. J Immunol 131:633–637

Mond JJ, Lees A, Snapper CM (1995) T cell-independent antigens type 2. Ann Rev Immunol 13:655–692

Mosier DE, Mond JJ, Goldings EA (1977) The ontogeny of thymic independent antibody responses in vitro in normal mice and mice with X-linked B cell defect. J Immunol 119:1874–1878

Neuhaus FC, Baddiley J (2003) A continuum of anionic charge: structures and functions of D-alanyl-teichoic acids in Gram-positive bacteria. Microbiol Molec Biol Rev 67:686–723

Niiro H, Clark EA (2002) Regulation of B-cell fate by antigen-receptor signals. Nat Rev Immunol 2:945–956

Peeters CC, Tenbergen-Meekes AM, Poolman JT, Beurret M, Zegers BJ, Rikers GT (1991) Effect of carrier priming on immunogenicity of saccharide-protein conjugate vaccines. Infect Immun 59:3504–3510

Poolman JT, Bakaletz L, Cripps A, Denoel PA, Forsgren A, Kyd J, Lobet Y (2000) Developing a nontypeable *Haemophilus influenzae* (NTHi) vaccine. Vaccine 19:S108–S115

Pozdnyakova O, Guttormsen HK, Lalani FN, Carroll MC, Kasper DL (2003) Impaired antibody response to group B streptococcal type III capsular polysaccharide in C3- and complement receptor 2-deficient mice. J Immunol 170:84–90

Region-of-Waterloo-Public-Health (2004) Fact sheet for Menjugate® (meningococcal-C vaccine). DOCS, 152595

Riordan KO, Lee JC (2004) *Staphylococcus aureus* capsular polysaccharide. Clin Microbiol Rev 17:218–234

Robbins JB, Schneerson R (1990) Evaluating the *Haemophilus influenzae* type b conjugate vaccine PRP-D. N Engl J Med 323:1415–1416

Ruggeberg JU, Pollard AJ (2004) Meningococcal vaccines. Pediatr Drugs 6:251–266

Sanofi-Pasteur (2005a) Haemophilus b conjugate vaccine ActHIB®. Product insert, 095 3105021

Sanofi-Pasteur (2005b) Typhim Vi®, Vi capsular polysaccharide typhoid vaccine. Patient information leaflet, 17387

Sanofi-Pasteur (2006) Meningococcal (groups A, C, Y and W-135) polysaccharide diphtheria toxoid conjugate vaccine, Menactra®. Prescribing information, 284 3108234

Sater RA, Sandel PC, Monroe JG (1998) B cell receptor-induced apoptosis in primary transitional murine B cells: signaling requirements and modulation by T cell help. Int Immunol 10:1673–1682

Scher I, Steinberg AD, Berning AK, Paul WE (1975) X-linked B-lymphocyte immune defect in CBA/N mice. II. Studies of the mechanisms underlying the immune defect. J Exp Med 142:637–650

Schwandner R, Dziarski R, Wesche H, Rothe M, Kirschning CJ (1999) Peptidoglycan- and lipoteichoic acid-induced cell activation is mediated by toll-like receptor 2. J Biol Chem 274:17406–17409

Sen G, Khan AQ, Chen Q, Snapper CM (2005) In vivo humoral immune responses to isolated pneumococcal polysaccharides are dependent on the presence of associated TLR ligands. J Immunol 175:3084–3091

Sidman CL, Unanue ER (1975) Receptor-mediated inactivation of early B lymphocytes. Nature 257:149–151

Snapper CM, Mond JJ (1996) A model for induction of T cell-independent humoral immunity in response to polysaccharide antigens. J Immunol 157:2229–2233

Snapper CM, Shen Y, Khan AQ, Colino J, Zelazowski P, Mond JJ, Gause WC, Wu ZQ (2001) Distinct types of T-cell help for the induction of a humoral immune response to *Streptococcus pneumoniae*. Trends Immunol 22:308–311

Sproul TW, Malapati S, Kim J, Pierce SK (2000) Cutting edge: B cell antigen receptor signaling occurs outside lipid rafts in immature B cells. J Immunol 165:6020–6023

Sverremark E, Fernandez C (1998) Role of T cells and germinal center formation in the generation of immune responses to the thymus-independent carbohydrate dextran B512. J Immunol 161:4646–4651

Treiner E, Lantz O (2006) CD1d- and MR1-resticted invariant T cells: of mice and men. Curr Opin Immunol 18:519–526

Vos Q, Lees A, Wu ZQ, Snapper CM, Mond JJ (2000) B-cell activation by T-cell-independent type 2 antigens as an integral part of the humoral immune response to pathogenic microorganisms. Immunol Rev 176:154–170

Weidenmaier C, Kokai-Kun JF, Kristian SA, Chanturiya T, Kalbacher H, Gross M, Nicholson G, Neumeister B, Mond JJ, Peschel A (2004) Role of teichoic acids in *Staphylococcus aureus* nasal colonization, a major risk factor in nosocomial infections. Nature Med 10:243–245

Weintraub BC, Jun JE, Bishop AC, Shokat KM, Thomas ML, Goodnow CC (2000) Entry of B cell receptor into signaling domains is inhibited in tolerant B cells. J Exp Med 191:1443–1448

Whitney CG (2005) Impact of conjugate pneumococcal vaccines. Pediatr Infect Dis J 24:729–730

Wu ZQ, Vos Q, Shen Y, Lees A, Wilson SR, Briles DE, Gause WC, Mond JJ, Snapper CM (1999) In vivo polysaccharide-specific IgG isotype responses to intact *Streptococcus pneumoniae* are T cell dependent and require CD40- and B7-ligand interactions. J Immunol 163:659–667

Wu ZQ, Khan AQ, Shen Y, Schartman J, Peach R, Lees A, Mond JJ, Gause WC, Snapper CM (2000) B7 requirements for primary and secondary protein- and polysaccharide-specific Ig isotype responses to *Streptococcus pneumoniae*. J Immunol 165:6840–3848

Wyeth (2005) Haemophilus b conjugate vaccine HibTITER®. Product insert, W10461C004

Wyeth (2006) Pneumococcal 7-valent conjugate vaccine (diphtheria CRM 197 protein) Prevnar®. Prescribing information, W10430C006

Wyeth-Australia (2005) Meningitec® Meningococcal Group C Conjugate Vaccine. Consumer medicine information, AUST R 75721

Zarember KA, Godowski PJ (2002) Tissue expression of human TLR and differential regulation of TLR mRNA's in leukocytes in response to microbes, their products, and cytokines. J Immunol 168:554–561

B Cell Lineage Contributions to Antiviral Host Responses

N. Baumgarth(✉), Y. S. Choi, K. Rothaeusler,
Y. Yang, and L. A. Herzenberg

1 Introduction 42
2 Multiple B Cell Subsets Contribute to Humoral Immune Defenses 43
   2.1 Follicular B Cells 43
   2.2 Marginal Zone B Cells 44
   2.3 B-1 Cells 46
3 Immunity to Influenza Virus Infection 51
   3.1 Follicular B Cell Responses in Influenza Virus Infection 52
   3.2 MZ B Cells and Influenza Virus Infection 53
   3.3 B-1 Cell Contributions to Influenza Virus-Specific Immunity 54
References 57

Abstract B cell responses are a major immune protective mechanism induced against a large variety of pathogens. Technical advances over the last decade, particularly in the isolation and characterization of B cell subsets by multicolor flow cytometry, have demonstrated the multifaceted nature of pathogen-induced B cell responses. In addition to participation by the major follicular B cell population, three B cell subsets are now recognized as key contributors to pathogen-induced host defenses: marginal zone (MZ) B cells, B-1a and B-1b cells. Each of these subsets seems to require unique activation signals and to react with distinct response patterns. Here we provide a brief review of the main developmental and functional features of these B cell subsets. Furthermore, we outline our current understanding of how each subset contributes to the humoral response to influenza virus infection and what regulates their differential responses. Understanding of the multilayered nature of the humoral responses to infectious agents and the complex innate immune signals that shape pathogen-specific humoral responses are likely at the heart of enhancing our ability to induce appropriate and long-lasting humoral responses for prophylaxis and therapy.

N. Baumgarth
Center for Comparative Medicine, University of California, Davis,
County Rd 98 & Hutchison Drive, Davis, CA 95616, USA
e-mail: nbaumgarth@ucdavis.edu

Abbreviations BCR: B cell receptor; d.p.c: Days postconception; MZ: Marginal zone; TNP: Trinitrophenyl

1 Introduction

B cell responses are a major immune protective mechanism induced against a large variety of pathogens. A number of antibody effector mechanisms provide immune protection. For example, the direct binding of antibodies can interfere with pathogen attachment to host cells, thereby inhibiting cell entry and replication of intracellular pathogens. In addition, antibody binding can activate complement for direct destruction of pathogens, or interfere with other less well-understood mechanisms that inhibit infectivity of a pathogen.

The induction of B cell responses has been successfully exploited as the protective immune mechanism induced by most currently available vaccines, including inactivated influenza virus vaccines. Nonetheless, vaccine-induced responses are often of smaller magnitude and shorter duration compared to those induced to live-pathogen infection. While the underlying mechanisms for these differences are poorly understood, it is likely that the induction of innate regulatory mechanisms triggered specifically to certain classes of pathogens is at the base for some of those observed differences.

In addition to direct antibody-mediated pathogen neutralization, antibodies play a significant role in enabling or enhancing immune defenses. Importantly, many of these mechanisms have been associated with innate immune protection, which occurs very early after infection. Antibody-enhanced innate immune functions extend from activation of the classical pathway of complement by antibody–antigen complexes to antibody-dependent cellular cytotoxicity by natural killer cells. Table 1 provides a list of some protective mechanisms affected by the presence of antibodies. Given the low frequency of antigen-specific B cells at the onset of an infection and a general requirement for simultaneously induced T cell help, these findings raise the question of whether induction of early antibody responses underlies the same regulatory mechanisms as those induced later and those that form memory responses.

In this review, we will discuss the multifaceted nature of pathogen-induced B cell responses elaborated by the activation of distinct B cell subsets with unique functional characteristics and emphasize the role of B-1 cells in this process. The concur-

Table 1 Antibody mediated immune protection during early infection

- Direct pathogen neutralization/inactivation
- Pathogen opsonization for uptake by macrophages and dendritic cells
- Natural killer cell activation and antibody-mediated cellular cytotoxicity
- Activation of the classical pathway of complement
- B cell activation via positive feedback stimulation through complement receptors
- Formation of antigen-antibody complexes for antigen-presentation by follicular dendritic cells

rent activation of various B cell subsets forms the basis for the large contribution of B cells in protection from both acute and recurrent infections. We will summarize our recent findings with regard to the antibody responses induced to live influenza virus infection and the simulation of bacterial infection via injection with LPS.

2 Multiple B Cell Subsets Contribute to Humoral Immune Defenses

Technical advances over the last decade, particularly in the isolation and characterization of B cell subsets by multicolor flow cytometry, have demonstrated the multifaceted nature of pathogen-induced B cell responses. It is now understood that in addition to the majority follicular B cell population, other B cell subsets contribute to many pathogen-specific humoral responses. Those additional B cell subsets are identified as marginal zone (MZ) B cells, B-1a and B-1b cells. Each B cell subset seems to require unique activation signals and to respond with distinct response patterns. Whether the follicular B cell population itself contains further subsets with differing activation requirements and/or functions is unknown. Current identification of B cell subsets is based mainly on cell surface phenotypic analysis, which identifies follicular B cells as a very homogeneous lymphocyte population.

2.1 Follicular B Cells

2.1.1 Development

Most B cells reside in the follicles of secondary lymphoid tissues of adult humans and rodents. Their developmental pathway from an uncommitted hematopoietic stem cell to an immature B cell in the bone marrow has been extensively characterized over the last 20 years (Hardy and Hayakawa 2001; Meffre et al. 2000; Rolink et al. 2001). The precise mechanisms that cause further differentiation or the death of immature B cells, once they have reached the spleen as so-called transitional B cells, are incompletely understood. Survival depends at minimum on the continuous presence of a BCR (Kraus et al. 2004) that lacks high-affinity/specificity to self-antigens (Goodnow et al. 1995). Follicular B cells have a half-life of 4–5 months in mice (Forster and Rajewsky 1990) and are continuously replenished from bone marrow precursors.

2.1.2 Function

Follicular B cells, also termed conventional or B-2 cells, generate the bulk of the induced antibody responses following protein immunization. Responses are generated following establishment of vigorous germinal centers, seeded initially by follicular

B cells of relative low affinity for antigen (Paus et al. 2006). Much has been learned about the molecular mechanisms underlying conventional B cell response induction. In general, the responses are dependent on a minimum of two signals: B cell receptor (BCR) signaling via antigen binding and CD40–CD40L-mediated T cell help (Bernard et al. 2005; Noelle 1996). Thus, significant conventional B cell contributions require antigen-stimulated T cells, which are rare at the onset of an infection. These responses require some time to develop in order for clonal expansion of both T and B cells to occur and for T–B interaction to take place (Baumgarth 2000). Thus, T-dependent conventional B cell responses cannot be responsible for all rapid antibody responses generated to infections. Furthermore, recent literature suggests that innate third signals further enhance and modulate these responses. Two such signals have been identified to date as stimulation through TLR, via a MyD88-dependent pathway (Pasare and Medzhitov 2005; Ruprecht and Lanzavecchia 2006) and stimulation of B cells via type I (alpha and beta) interferon (Chang et al. 2007; Coro et al. 2006; Fink et al. 2006; Le Bon et al. 2006).

T-independent responses can be elaborated by follicular B cells and might explain in part rapid response induction to pathogens. Pathogens often express strongly structured, repetitive units on their surface. In the experimental setting, these are mimicked by haptens bound to a carrier backbone, creating multivalent binding sites (Bachmann and Zinkernagel 1997; see also the chapter by H.J. Hinton et al., this volume). Early studies on responses to the hapten trinitrophenyl (TNP) indicated that TNP-specific B cell responses to TNP-LPS are contributed by the same B cell precursors as those contributed by B cell responding in a T-dependent fashion (to TNP–red blood cell conjugates) (Lewis et al. 1978). Thus, TLR-mediated B cell stimulation, at a minimum stimulation through TLR4, might act to facilitate enhanced follicular B cell activation early during a response by providing directly and indirectly additional signals to B cells that can replace the need for T cell help. These third signals also support enhanced responses when T cell help is present (Pasare and Medzhitov 2005; Ruprecht and Lanzavecchia 2006).

2.2 *Marginal Zone B Cells*

2.2.1 Development

In the adult, MZ B cells and follicular B cells seem to develop from a common B cell precursor in the bone marrow. As their name suggests, following bone marrow development these B cells accumulate in the marginal zone of the spleen (Pillai et al. 2005). Establishment of the splenic marginal zone B cell compartment is slow, however, requiring some weeks to establish following birth or after whole body irradiation in experimental mouse models. In addition to these bone marrow-derived MZ B cells, there is evidence from studies in various genetically manipulated mice that MZ B cells are also contributed from precursors in fetal liver and spleen during the pre- and neonatal period (Carvalho et al. 2001; Hao and Rajewsky

2001; Heltemes-Harris et al. 2005). Ablation of de novo B cell development in adult mice via inducible deletion of the BCR causes a reduction of follicular B cells over time, while stable numbers of MZ or B-1 cells are maintained (Hao and Rajewsky 2001). Together, the data suggest that MZ B cells have significantly longer half-lives compared to follicular B cells and/or that their developmental origin is distinct from that of follicular B cells. Recent findings from a number of gene-targeted mice lacking various BCR-signaling components and transcription factors further indicate that MZ and follicular B cells require distinct signals for their development (Pillai et al. 2005). Importantly, exogenous antigen does not seem to be required for the selection of B cells into the marginal zone pool (Dammers and Kroese 2005). Phenotypically, MZ B cells are differentiated from follicular B cells by their high surface expression of the complement receptor CD21, CD9, low expression of IgD and lack of CD23.

2.2.2 Function

MZ B cells represent a B cell subpopulation uniquely positioned to provide rapid responses (Martin et al. 2001). Their location at the marginal zone of the spleen provides them with rapid exposure to blood-borne antigens. It also places them in close proximity to marginal zone macrophages that might present antigen and provide potential other innate stimuli. One consequence of splenectomy in humans and in experimental animal models is an increased risk of bacterial sepsis.

MZ B cells have a higher propensity to respond more rapidly and vigorously to innate signals such as LPS compared to follicular B cells (Martin et al. 2001). This is a reflection of their genetic make-up, which includes higher expression of various integrins and innate receptors compared to follicular B cells (Lopes-Carvalho et al. 2005). They also express higher levels of costimulatory molecules such as CD80/CD86 and in contrast to naïve follicular B cells are able to prime naïve T cells (Attanavanich and Kearney 2004). MZ B cell activation results in the establishment of vigorous extrafollicular foci responses, leading to strong early antibody secretion. While they can form germinal centers, those responses usually occur late in the response and are less prominent (Song and Cerny 2003). While MZ B cells might not generate vigorous germinal center responses, their early extrafollicular secretion of IgM does facilitate the establishment of germinal centers. The secretion of IgM by MZ B cells was shown to enable the complement receptor-dependent deposition of IgM-antigen complexes on the surface of follicular B cells (Ferguson et al. 2004). This is in agreement with studies in IgM secretion-deficient mice, which demonstrated the need for IgM in the establishment of strong IgG responses (Baumgarth et al. 2000b; Boes et al. 1998a).

The distinct tissue architecture of the spleen and the flow of blood that pools initially in the marginal zone sinuses make the spleen a unique and potent blood-filtering organ. Thus, splenic marginal zone B cells participate effectively in response to blood-borne and/or systemic infections. However, such infections are relatively rare, as the vast majority of infections occur via mucosal surfaces of the gastrointestinal and respiratory tracts. Rapid local immune responses might thus not benefit from the presence of this cell population, since they occur primarily in the regional lymph nodes where afferent lymphatic enter the lymph nodes through the cortical sinuses. Lymph nodes lack phenotypic marginal zone B cell equivalents, i.e., $CD21^{hi}$ $CD23^{neg}$ IgM^{hi} IgD^{lo} cells. B cells in lymph nodes from noninfected or inflamed tissue sites appear very homogenous in flow cytometric analysis. Virtually all lymph node B cells are classical $CD21^{int.}$ $CD23^+$ IgM^{lo} IgD^{hi} follicular B cells. Ongoing studies suggest, however, that functional equivalents of splenic MZ B cells might exist (Rothaeusler and Baumgarth, see Sect. 3.2).

2.3 B-1 Cells

2.3.1 Development

The development of B-1a ($CD5^+$) and B-1b ($CD5^-$) cells is still incompletely understood and subject of some controversy. B-1 cells are the first B cells to develop in ontogeny (Kantor and Herzenberg 1993). The first (B-1 cell-restricted) B cell precursors are found in the splanchnopleure of the developing mouse embryo, approximately 7–9 days postconception (d.p.c.) (Godin et al. 1993). B-1 cells are also the main B cell population emerging from the fetal liver starting from around day 12 d.p.c. (Hardy and Hayakawa 1991).

Development of B-1 cells from dedicated bone marrow precursors after weaning is rare compared to B-2 cell development. While initial adoptive transfer studies suggested a complete absence of B-1 cell development from the bone marrow (Hardy and Hayakawa 1991; Hayakawa et al. 1985), our later studies demonstrated that adult bone marrow commonly reconstitutes a small population of B-1a cells and nearly half of the B-1b cells in irradiated adoptive recipients (Kantor and Herzenberg 1993; Kantor et al. 1995). More recently, studies with purified hematopoietic stem cells (Kikuchi and Kondo 2006) and B cell precursors (Montecino-Rodriguez et al. 2006) indicate that bone marrow contains a rare B-1 restricted precursor that can provide limited B-1 reconstitution. Findings from these latter studies (Montecino-Rodriguez et al. 2006) strongly support our long-held view that B-1 and B-2 cells belong to distinct developmental lineages (Herzenberg et al. 1986, 1992; Herzenberg and Tung 2006; Kantor and Herzenberg 1993).

Several studies suggest that B-1a and B-1b belong to separate developmental lineages (Herzenberg et al. 1992; Herzenberg and Tung 2006; Kantor and Herzenberg 1993; Stall et al. 1992). However, the question of whether B-1a progenitors, or B-1a cells themselves, can give rise to B-1b cells is still open. As in earlier adoptive transfer studies, we find that FACS-sorted $CD5^+$ B-1a cells commonly reconstitute a portion of the peritoneal ($CD5^-$) B-1b compartment (N. Baumgarth and L.A. Herzenberg, unpublished observations). Published studies by others also support the finding that transfer of FACS-purified $CD5^+$ B-1 cells can result in reconstitution of some B-1b cells (Haas et al. 2005). The interpretation of this finding is unclear since there is substantial overlap between the CD5⁺ and CD5⁻ B-1 populations in the peritoneal cavity and neither we nor others can rule out the possibility that contaminating B-1b cells are responsible for the B-1b reconstitution observed with the sorted B-1a cells. Nevertheless, we should not ignore the possibility that at least some B cells, phenotypically classified as B-1b cells, are in fact (activated?) B-1a cells with downregulated CD5 expression.

The identification of B-1 cells and their separation into B-1a and B-1b sister populations (Stall et al. 1992) was done with body-cavity B cells (pleural and peritoneal). However, even within the peritoneal cavity B-1 cells are not homogenous in their expression of often used B-1 cell markers. For example, both CD11b and CD43 are not expressed on all peritoneal cavity B cells that otherwise show hallmarks of a B-1 cells ($B220^{lo}$ IgM^{hi} IgD^{lo}). Whether B-1 cells that differ in their expression of CD11b and CD43 have different functional properties remains to be established. Identification of B cell subsets with the help of these surface markers is also insufficient because of its limitations to the analysis of body-cavity B cells. Splenic B-1 cells do not express CD11b (nor do all cavity B-1 cells) and no other markers are known to distinguish the B-1b cells from activated B cells. For example, low expression of B220 plus expression of CD43, another characteristic staining pattern of B-1 cells (Wells et al. 1994), is also found among activated B cell blasts (reviewed in: Baumgarth 2004). Furthermore, certain stimulation conditions will induce the expression of CD5 on B-2 cells in vitro (Berland and Wortis 2002). In BCR-transgenic mice carrying a B cell self-antigen (hen egg lysozyme), developing anergic B cells were also shown to express CD5 (Hippen et al. 2000). Thus, clear identification of B-1a and B-1b cells outside the body cavities in nonmanipulated mice and consequently in humans, is exceedingly difficult. In spleen, it is far more difficult to resolve B-1a from B-1b since B-1b only represent 5%–10% of total B-1 cells and total B-1 in spleen in unstimulated animals usually only represent less than 1%–2% of total B cells. Nevertheless, sensitive FACS assays reveal the presence of B-1a and B-1b in both spleen and PerC (Fig. 1).

2.3.2 Function

Although B-1 cells are present at relatively low frequencies in secondary lymphoid organs, they are well established as the producers of much of the circulating natural antibody in both humans and mice (Baumgarth et al. 1999). The advantages of such evolutionary B cell memory on the survival of a particular species are obvious (Baumgarth et al. 2005). It is therefore not surprising that common antigenic patterns are often the target of natural antibodies. For example, the ubiquitous presence in mice of natural antibodies to phosphorylcholine, a Gram-positive bacterial cell wall component, clearly suggests that shaping of the natural antibody repertoire by evolutionary pressures has resulted in key protection against bacterial pathogens. How such preferential rearrangement, or the selective expansion of B-1 cells exhibiting certain specificities, is achieved might be at the heart of understanding the development of B-1 cells.

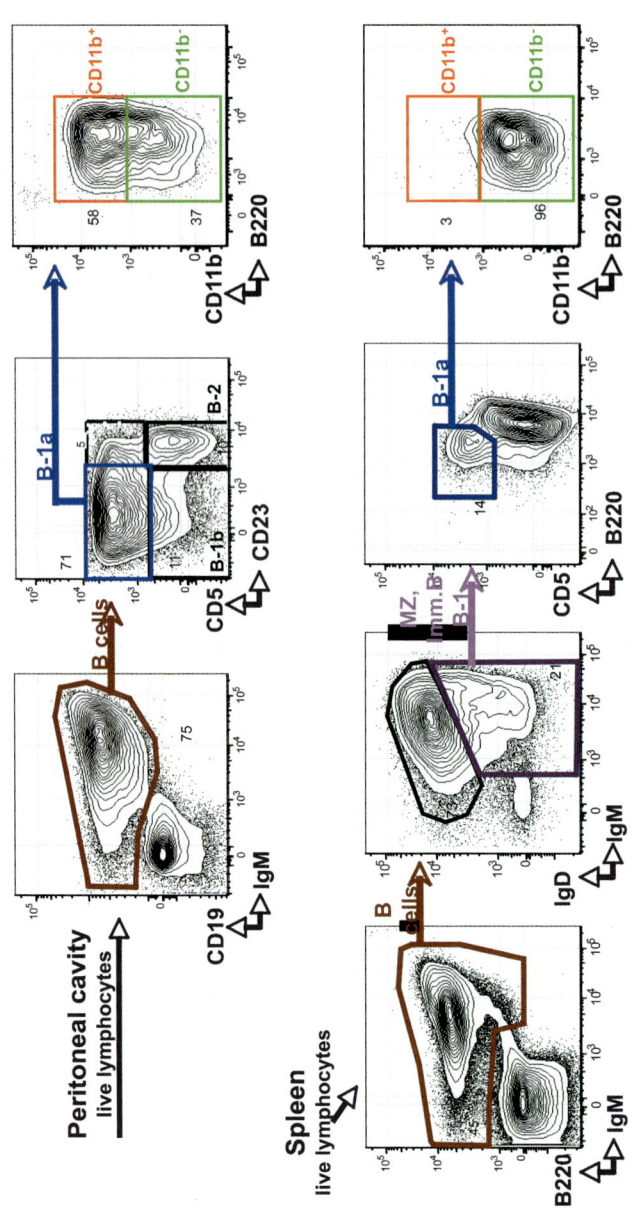

Fig. 1 Peritoneal cavity and splenic B-1a cells. Shown are 5% contour plot with outliers of peritoneal cavity (*top*) and splenic (*bottom*) cell suspensions. As shown, the majority of peritoneal cavity B-1a cells (CD19⁺ IgM⁺ CD5⁺ CD23⁻ B cells) express CD11b. In contrast, most splenic B-1a cells (CD5⁺B220^{lo}IgM^{hi}IgD^{lo}) lack CD11b expression

B-1 cells are typically thought of as peritoneal and pleural B cells because they are the predominant B cell populations at these locations. However, it is important to recognize that despite their low frequency (~1%), substantial numbers of B-1 cells are found in the spleen. In fact, the total number of B-1 cells in the organ in unstimulated animals (~10^6) is only slightly lower than the number of B-1 cells in PerC (~2×10^6). Furthermore, after LPS stimulation, the number of B-1 cells in the spleen parallels the increase in total cellularity, resulting in spleens that contain roughly 4×10^6 B-1 cells several days after stimulation, and hence a substantially larger number of B-1 cells than is typically found in PerC in unstimulated animals.

Consistent with the demonstration that B-1 cells disappear rapidly from the peritoneal cavity after intraperitoneal LPS stimulation (Ha et al. 2006), we have recently shown that the increase in B-1 cells after intravenous LPS stimulation is due to rapid migration and division of PerC B-1 cells into the spleen (Yang et al. 2007). These immigrants are identifiable for the first few days after they arrive in the spleen because they express CD11b at the same level as expressed on most PerC B-1 (Fig. 2). However, after 6 days, they lose CD11b expression and can only be tracked by experimentally introduced markers (e.g., Ig allotype). At least 30% of the immigrant B-1 cells in spleen divide within 1 day of LPS-stimulated animals. Within 2 days, roughly 10% of the cells that divided differentiate to become mature antibody-secreting plasma cells (CD138^{hi} Blimp-1^{hi} intracellular IgM$^+$). In contrast, minimal cell division and no detectable plasma cell development occur in PerC in the LPS-stimulated animals.

Importantly, there is an initial wave of plasma cell development that occurs in the absence of cell division (Yang et al. 2007). These plasma cells are all derived from resident B-1 cells that have terminated CD11b expression. Only a proportion of the resident B-1 cells participate in this initial wave, which enables the appearance of mature plasma cells 1–1.5 days after LPS stimulation (Fig. 2). The immigrant B-1 cells, in contrast, do not begin to reach the mature pool until 2 days after LPS. Thus, the first wave of innate antibodies produced in the spleen is selectively derived from a unique B-1 population that resides in the spleen and is capable of rapid differentiation in the absence of division.

Initial examination of the antibody repertoire expressed by the plasma cells that develop during the first wave of the response to LPS stimulation indicates that their repertoire is enriched for cells expressing common natural antibodies produced by B-1 cells (e.g., VH11Vk9, Y. Yang and L.A. Herzenberg, unpublished observations). In essence, the frequency of cells expressing this antibody is roughly three- to fourfold higher among these plasma cells than among the B-1 plasma cells that develop 2–3 days later. Thus, evolution appears to have devised a mechanism that places these well-known B-1 antibodies in a position to be the first to be produced when a bacterial stimulus such as LPS is encountered.

In addition to the important contributions of B-1 cells in providing protective natural antibodies to both bacterial and viral pathogens (Baumgarth et al. 2000a; Boes et al. 1998b; Ochsenbein et al. 1999), B-1 cells can also actively participate in the induction of at least some immune responses. Recent studies by Tedder's group suggest a division of labor by which B-1a cells provide natural antibodies

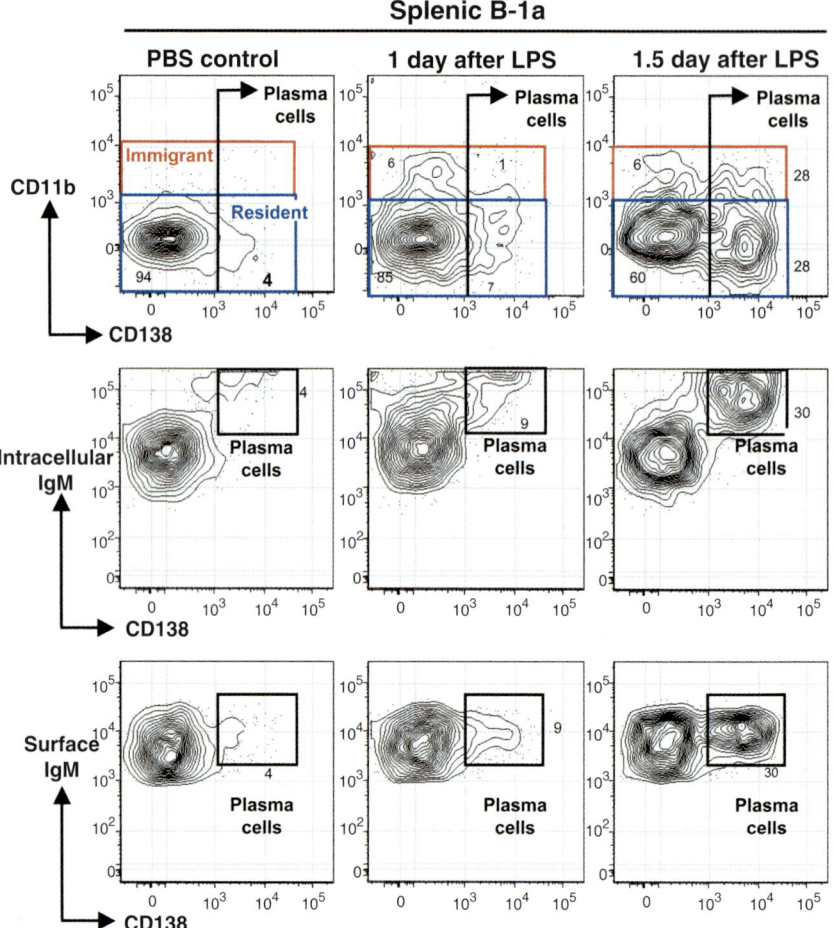

Fig. 2 Stimulation of peritoneal cavity B-1 cells with LPS triggers their migration to the spleen. Shown are 5% contour plots with outliers of splenic B cells gated as shown in Fig. 1. Cells were isolated from the spleen of mice injected with PBS (*left*) or LPS at 24 h (*middle*) and 36 h (*right panel*) prior to analysis. Note that new immigrants are distinct from resident B-1 cells by their expression of CD11b. Differentiation of resident B-1 cells to plasma cells is indicated by their upregulation of CD138 (syndecan-1) and high expression of intracellular IgM (*middle panel*), while maintaining surface IgM (*lower panel*)

and B-1b are induced to respond to *Streptococcus pneumoniae* infection with enhanced antibody secretion (Haas et al. 2005). This induced response seems to provide much of the polysaccharide-specific antibodies. Given the strong inhibitory activity of CD5 on BCR signaling (Bikah et al. 1996), these data provide an appealing explanation for the regulation of specific B-1 cell responses: B-1 cell responses would be restricted to those that do not express the inhibitory CD5 receptor.

In support of this view, studies with the relapsing fever-inducing spirochete *Borrelia hermsii* showed that protective activity could be transferred with IgM-secreting B-1b cells (Alugupalli et al. 2004; see also the chapter by K.R. Alugupalli, this volume) and that the B-1b cells in situ are sufficient for protection against the parasite (another example of B-1b cell responses). On the other hand, early work by Kenny and colleagues clearly established that B-1a cells (CD5$^+$) uniquely express T15-idiotype antibodies that react specifically with phosphorylcholine (Kenny et al. 1983; Knoops et al. 2004). IgM antibodies produced by these B cells are strongly increased in response to *S. pneumoniae* infection and contribute significantly to the primary response to the bacterium (Kenny et al. 1983). In addition, Peter Ernst and colleagues have shown that B-1a cells producing T15 idiotype antibodies are required for protection against mucosal (gut) infection (Pecquet et al. 1992a, 1992b). Thus, at least some B-1a cells contribute actively to protection against bacterial infection, raising doubt about the general applicability of the idea that B-1a play only a passive role, while B-1b cells play an active role immune responses (Kawikova et al. 2004; Pecquet et al. 1992a; Szczepanik et al. 2003).

3 Immunity to Influenza Virus Infection

Influenza's main evasion strategy relies on rapid replication and aerosol-mediated viral spread. Indeed, the relatively small influenza virus genome seems to contain only one gene (NS-1) involved in immune evasion strategies. NS-1 acts via binding to viral genomic single-stranded RNA, thereby inhibiting activation of the RNA helicase enzymes retinoic acid-inducible gene I (RIG-1) and the induction of type I interferon (Pichlmair et al. 2006). Infection with influenza virus induces an array of cellular and humoral immune defense mechanisms, both innate (Fujisawa et al. 1987; Reading et al. 1997) and adoptive (Doherty 2000; Doherty et al. 1997; Gerhard et al. 1997), which act in concert to provide strong protective immunity. Natural killer cells, macrophages (Fujisawa et al. 1987), natural antibodies (Baumgarth et al. 2000a), and the induction of type I interferon (Basler et al. 2001; Garcia-Sastre et al. 1998; Talon et al. 2000; Wang et al. 2000) and IL-1 (Schmitz et al. 2005) provide a first line of immune defense. Strongly cytolytic, virus-specific CD8$^+$ T cells (Doherty et al. 1997; Flynn et al. 1998; Hogan et al. 2001) and neutralizing antibodies (Gerhard et al. 1997) provide an effective way for removing virus-infected host cells and inactivating infectious virus. The development of humoral responses is of great importance, as they can provide disease-preventing sterile immunity through local production of antibodies (Renegar and Small 1991a, 1991b) and their induction through vaccination is currently used as an effective means of protection (Bridges et al. 2001).

The humoral response to influenza virus is comprised of different sets of antibodies: (a) natural antibodies, produced prior to any encounter with the virus (Baumgarth et al. 1999, 2000b), (b) virus-induced antibodies produced in a T cell-dependent manner, and (c) virus-induced antibodies produced independent of cognate T cell help (Gerhard et al. 1997; Mozdzanowska et al. 1997, 2000; Sha and Compans 2000; Virelizier et al. 1974). These latter types of antibodies provide a strong component of the immune response. As we have shown previously, natural antibodies that bind influenza virus are crucial for survival from infection (Baumgarth et al. 2000a). Gerhard and colleagues showed that T-independent virus-induced B cell responses provide immune protection against influenza virus (Mozdzanowska et al. 2000). Thus, both natural antibodies, produced prior to encounter with pathogens and T cell-independent pathogen-induced antibodies are being increasingly recognized as important components of the humoral response. Understanding the contribution of distinct B cell subsets might help to determine the mechanisms of these unconventional responses.

3.1 Follicular B Cell Responses in Influenza Virus Infection

Follicular B cells generate the majority of the influenza-virus induced antibody responses. Much of this response is T-dependent, shown by the massive reduction in antibody levels in T-deficient or CD4 T cell-deficient or T-deficient nude mice. Following experimental infection of mice, germinal center responses develop in the draining mediastinal lymph nodes around day 6–7 of infection (N. Baumgarth and K. Rothaeusler, unpublished observations). At this time, systemic antibody levels in the serum are also detectable for the first time (Baumgarth et al. 1999; Baumgarth and Kelso 1996).

The local induction of virus-specific conventional B-2 cell responses in the draining lymph nodes seem uniquely affected by the presence of inflammatory cytokines. Studies on IL-1R1 gene-targeted mice showed an enhancing effect of virus-induced IL-1 production on IgM responses (Schmitz et al. 2005). Our recent studies demonstrated the type I interferon-dependent activation of all lymph node B cells early during influenza virus infection (Chen et al. 2007; Coro et al. 2006). This innate early B cell activation was not only required for strong antibody responses to the virus (Coro et al. 2006), it also had significant consequences for the response outcome following BCR-mediated stimulation (Chen et al. 2007). Interestingly, among the strongest infection-induced gene expression changes in regional B cells were the upregulation of TLR3 and TLR7 (Chen et al. 2007). TLR7 and type I IFN are important regulators of the isotype profile of the developing antiviral response to influenza (Coro et al. 2006; Heer et al. 2007). Thus, infection-mediated innate B cell stimulation alters the way in which B cells respond to both antigen and innate signals. How both specific BCR-mediated and nonspecific TLR-mediated signals are synthesized to optimize B cell response outcomes is an important question requiring further study. B cell response model systems in which B cell responses have been studied and which have formed the basis for our understanding of B cell response regulation do not provide these innate signals and thus might not be sufficient to fully comprehend B cell response regulation to pathogen encounter. In addition, tissue-specific signals provided at the site of pathogen entry, most frequently the mucosal surfaces of the gastrointestinal or respiratory tract, might provide further signals distinctly regulating local compared to (splenic) systemic responses.

3.2 MZ B Cells and Influenza Virus Infection

Typical sublethal infection of mice (and humans) with influenza virus causes only localized respiratory tract infections, since the virus can only fully replicate in the respiratory tract epithelium. Therefore, the majority of viral antigen is not likely to enter the blood in any significant amount and marginal zone B cell responses would not be expected. However, ELISPOT analyses of murine spleen cells following influenza virus infection show a small but significant early and transient induction of virus-specific responses in the spleen (N. Baumgarth and K. Rothaeusler, unpublished observations). Flow cytometric analysis of spleens from influenza virus-infected mice demonstrates, however, a transient relative and absolute reduction in marginal zone B cells (Fig. 3). This latter finding is surprising, given that marginal

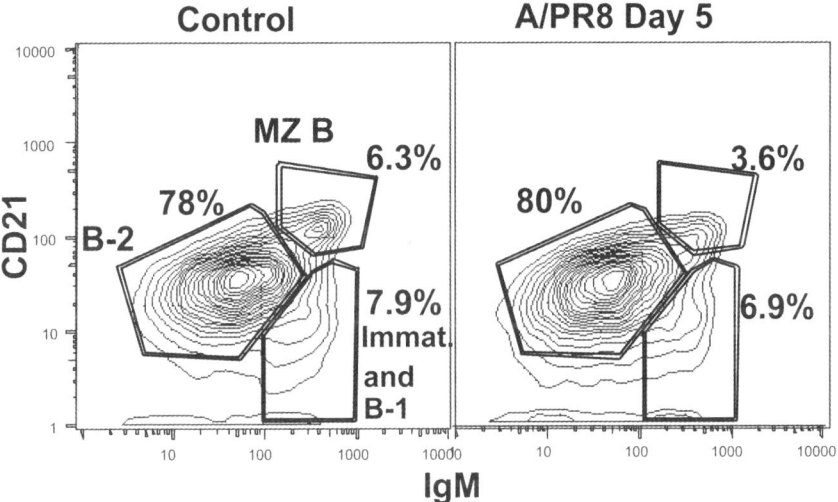

Fig. 3 Reduction in marginal zone B cells following acute influenza virus infection. Shown are 5% contour plots from splenic cells selected for expression of the pan-B cell marker B220 and lack of expression of CD3, 4,8, F4/80 and exclusion of propidium iodide as live/dead discriminator. Numbers indicate the relative proportion of marginal zone B cells (MZ), follicular (B-2), and immature/B-1 cells in the spleens of noninfected (*left panel*) and 5-day influenza A/PR8-infected mice (*right panel*). Note the strong reduction in MZ B cells expressing high levels of CD21 and IgM

zone B cells are regarded as sessile. While we have not identified the mechanism for this temporal depletion of the marginal zone B cell population, it indicates that marginal zone B cells can react to tissue-localized infections, possibly in response to circulating cytokines or chemokines. Given the need for sphingosine 1-phosphate receptor 1 expression in the appropriate localization of MZ B cells (Cinamon et al. 2004), infection-induced alterations in expression of this receptor or its ligand provides an attractive possible mechanism.

Could these data also indicate a mobilization of marginal zone B cells to the site of infection? As stated above, $CD21^{hi}$ MZ B cells are not present in the lymph nodes prior to infection and we did not find any $CD21^{hi}$ B cells in the regional mediastinal lymph nodes at any time after infection (K. Rothaeulser and N. Baumgarth, unpublished observations). While it is possible that phenotypically altered marginal zone B cells might accumulate in the lymph nodes following infection, existing data do not support this conclusion. MZ B cells do not alter their cell surface phenotype when they are dislodged from their proper location by genetic ablation of S1P1 or treatment with FTY720, at least within the spleen (Cinamon et al. 2004). A more likely explanation is the migration of activated marginal zone B cells into the red pulp, where they might reside short-term as differentiated plasma cells.

Given the fact that most infections occur localized and are contained within a certain tissue, it appears counterintuitive that populations of rapidly responding B cells are found only in the spleen but not in tissue-draining lymph nodes. In addition, we have provided evidence that following influenza virus infection isotype-switched antibody responses are measurable as early as day 2–3 after infection locally in the lymph nodes (Coro et al. 2006). Because of these very rapid kinetics, it appears safe to assume that these cells were activated by mechanisms other than cognate T cell help. Interestingly, earlier studies by Gerhard and colleagues demonstrated a strong difference in the idiotypic profile of the anti-influenza virus responses following immunization with influenza A/PR8. In particular, those studies identified a germline-encoded immunoglobulin-idiotype (C12Id) that contributed up to 25% of the entire early anti-hemagglutinin response but was absent from later primary and a secondary response (Kavaler et al. 1990, 1991). Our preliminary studies indicate that this response also dominates early B cell responses to infectious influenza virus. Furthermore, they suggest that the C12Id-response shows many of the functional hallmarks of a marginal zone B cell response, but it is contributed by cells that identify as follicular B cells (K. Rothaeusler and N. Baumgarth, unpublished observations). Thus, these data indicate that in lymph nodes rapidly responding (follicular) B cells exist that fulfill at least some of the roles MZ B cells play in the spleen.

3.3 B-1 Cell Contributions to Influenza Virus-Specific Immunity

The lack of unique markers to identify B-1 cells has hampered our understanding of their contributions to immunity against infections. In order to study the potential role of B-1 cells in influenza virus infection, we have therefore utilized protocols to

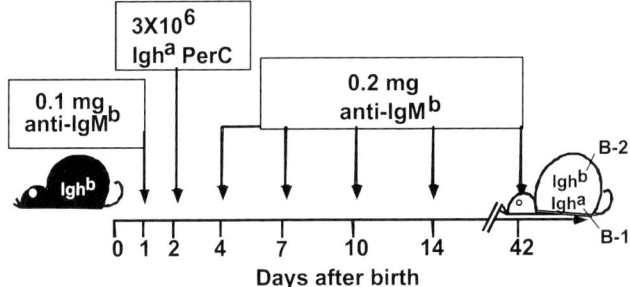

Fig. 4 Generation of B-1/B-2 allotype chimeric mice to track B-1 cell responses. Treatment of newborn mice with allotype-specific anti-IgM ablates all host-derived peripheral B cell development. This treatment is continued for 6 weeks. In this time, transferred congenic but allotype-disparate B-1 cells [or peritoneal cavity washout (PerC) cells as source of B-1 cells] expand. Four to 6 weeks following cessation of anti-IgM treatment, the host-derived B-2, but not the B-1 cell compartment is fully reconstituted. Donor-derived B-1 cells contribute >80% of the B-1 cell compartment of the mouse. (Baumgarth et al. 1999; Lalor et al. 1989a, 1989b)

track B-1 cells and their antibodies via immunoglobulin-allotype-specific markers (Fig. 4; Lalor et al. 1989a, 1989b). These mice are generated by treating newborn mice with host-allotype-specific anti-IgM on day 1 after birth and transferring congenic allotype-mismatched B-1 cells or peritoneal-cavity B cells on day 2 after birth. Four to six weeks after cessation of the anti-IgM treatment, host B-2 cells (but not host-B-1 cells) have returned to normal numbers. Most (>80%) B-1 cells in these mice are of the donor allotype, thus allowing tracking of donor-derived B-1 cells by their disparate Ig-allotype. While there are limitations to this approach, it provides a means of following B-1 cell responses without having to rely on potentially changing surface phenotypes.

3.3.1 Passive Contributions of Protection by Provision of Systemic Natural Antibodies

Using these allotype-chimeric mice, we demonstrated that mice harbor significant titers of natural, preinfection influenza-binding antibodies. These antibodies were generated mostly, if not exclusively, by B-1 cells (Baumgarth et al. 1999). Analysis of the serum antibody response to influenza virus further established that following infection the entire virus-induced antibody response, including virus-specific IgM, is derived from the host, i.e., the B-2 cell compartment. Thus, while B-1 cells seemed to generate virtually the entire natural antibody levels, they did not contribute to the systemic antibody response. Importantly, we showed that passive transfer of natural antibodies is at least partially protective and that the lack of IgM secretion by B-1 cells reduces survival from high-dose influenza virus infection (Baumgarth et al. 2000b). Thus, the generation of natural IgM importantly contributes to survival from an acute viral infection. Infections with other bacterial and viral pathogens have similarly shown the importance of this evolutionary conserved antibody-mediated immune defense mechanism (Boes et al. 1998b; Ochsenbein et al. 1999). Whether these antibodies were elaborated by B-1a or B-1b cells was not delineated in those studies.

It is important to note that B-1 cells contribute significantly to mucosal IgA production. Roughly one-third of plasma cells in lamina propria of B-1/B-2 allotype-chimeric mice are derived from B-1 cells (Kroese et al. 1989). We have made similar observations for the respiratory tract (N. Baumgarth and L.A. Herzenberg, unpublished observations). This is also in good agreement with recent studies by Stavnezer and colleagues showing that B-1 cells preferentially switch to IgA in vitro (Kaminski and Stavnezer 2006). Importantly, that study further confirms an earlier report (Tarlinton et al. 1995) that, while B-1 cells can undergo isotype-switching to all downstream isotypes, their responses are clearly distinct from that of B-2 cells, resulting mainly in IgM and to a lesser degree in IgA production.

3.3.2 Active Contributions of B-1 Cells in Local Respiratory Tract Immune Defense

Given the early literature on the specific T-15-idiotype encoded B-1a cell responses to S. *pneumoniae* infection outlined above (Kenny et al. 1983) and other studies that showed a strong induction of B-1 cell-derived auto-antibodies in a transgenic model following i.v. LPS injection (Murakami et al. 1994), our results were surprising to us at the time. More recent studies have also indicated the active involvement of B-1 cells, at least B-1b cells, in the systemic immune responses to *Borrelia hermsii* and *S. pneumoniae* infection (Alugupalli et al. 2004; Haas et al. 2005). Common to all published studies in which B-1 cell responses have been noted is the fact that the pathogen/LPS is administered systemically. In contrast, influenza virus infection is a localized infection of the respiratory tract. Therefore it appeared possible that B-1 cell responses to influenza virus infection might be induced in the respiratory tract but not systemically.

Ongoing studies are concerned with studying respiratory tract B-1 cell responses to influenza virus infection. Utilizing the same B-1/B-2 allotype-chimeric approach as in our previous studies, we now have experimental evidence that B-1 cells can indeed respond locally to influenza virus infection by accumulating in the draining lymph nodes and contributing to secretion of virus-binding and neutralizing IgM in the bronchoalveolar lavage (Y.S. Choi and N. Baumgarth, unpublished observations). Importantly, we have confirmed our original findings (Baumgarth et al. 1999) that systemic natural B-1 cell-derived antibody levels to influenza are unaffected, even in mice that have clear evidence of local B-1 cell responses. Thus, infection-induced local signals induce the accumulation and activation of B-1 cells in regional lymph nodes.

The mechanisms controlling B-1 cell migration following infections have not been studied in detail. CXCL13 was shown to be important for the homeostatic migration/accumulation of B-1 cells into the peritoneal cavity and B-1 cell responses to peritoneal streptococcal antigen immunization (Ansel et al. 2002). Studies by Fagarasan and colleagues (Ha et al. 2006) indicate a role for TLR-dependent integrin and CD9-mediated migration from the peritoneal cavity following LPS stimulation. However, from that published study it can also be concluded that most migration from the peritoneal cavity is not dependent on CD9/MyD88 as most B-1 cells seemed to have left the peritoneal cavity following adoptive transfer into $RAG^{-/-}$ mice within a few hours after transfer (Ha et al. 2006). Similarly, most B-1 cells rapidly vacate the pleural cavity following adoptive transfer into wild type mice (Y.S. Choi and N. Baumgarth, unpublished observations). Thus, the continuous migration of B-1 cells into and out of the body cavities is further enhanced by innate immune signals such as TLR-mediated stimuli provided during an infection. Whether this innate stimuli alone is sufficient for the subsequent accumulation of these cells in spleen or regional lymph nodes and their differentiation to antibody-secreting cells remains to be investigated. It also remains to be studied whether B-1 cell subsets show differences in their ability to migrate from the cavities to sites of infection.

In summary, protective humoral immunity to pathogens is contributed by distinct B cell subsets with unique activation requirements and response patterns. The nature and tissue distribution pattern of the pathogen strongly affects the quality of the induced response. This is due at least in part to the induction of innate immune signals that provide additional regulatory stimuli to B cell responses and by triggering distinct B cell subset responses. A better understanding of the individual B cell subset response components that contribute during a natural infection might enable us to design better vaccines, which induce appropriate, strong, and long-lasting humoral immune responses.

Acknowledgements N.B. acknowledges the financial support from NIH/NIAID (AI051354) and NIH training grant support to K.R. (T32-AI60555) for some of the outlined studies.

References

Alugupalli KR, Leong JM, Woodland RT, Muramatsu M, Honjo T, Gerstein RM (2004) B1b lymphocytes confer T cell-independent long-lasting immunity. Immunity 21:379–390

Ansel KM, Harris RB, Cyster JG (2002) CXCL13 is required for B1 cell homing, natural antibody production, and body cavity immunity. Immunity 16:67–76

Attanavanich K, Kearney JF (2004) Marginal zone, but not follicular B cells, are potent activators of naive CD4 T cells. J Immunol 172:803–811

Bachmann MF, Zinkernagel RM (1997) Neutralizing antiviral B cell responses. Annu Rev Immunol 15:235–270

Basler CF, Reid AH, Dybing JK, Janczewski TA, Fanning TG, Zheng H et al (2001) From the cover: sequence of the 1918 pandemic influenza virus nonstructural gene (NS) segment and characterization of recombinant viruses bearing the 1918 NS genes. Proc Natl Acad Sci U S A 98:2746–2751

Baumgarth N (2000) A two-phase model of B-cell activation. Immunol Rev 176:171–180

Baumgarth N (2004) B-cell immunophenotyping. Methods Cell Biol 75:643–662

Baumgarth N, Kelso A (1996) In vivo blockade of gamma interferon affects the influenza virus-induced humoral and the local cellular immune response in lung tissue. J Virol 70:4411–4418

Baumgarth N, Herman OC, Jager GC, Brown L, Herzenberg LA, Herzenberg LA (1999) Innate and acquired humoral immunities to influenza virus are mediated by distinct arms of the immune system. Proc Natl Acad Sci U S A 96:2250–2255

Baumgarth N, Chen J, Herman OC, Jager GC, Herzenberg LA (2000a) The role of B-1 and B-2 cells in immune protection from influenza virus infection. Curr Top Microbiol Immunol 252:163–169

Baumgarth N, Herman OC, Jager GC, Brown LE, Herzenberg LA, Chen J (2000b) B-1 and B-2 cell-derived immunoglobulin M antibodies are nonredundant components of the protective response to influenza virus infection. J Exp Med 192:271–280

Baumgarth N, Tung JW, Herzenberg LA (2005) Inherent specificities in natural antibodies: a key to immune defense against pathogen invasion. Springer Semin Immunopathol 26:347–362

Berland R, Wortis HH (2002) Origins and functions of B-1 cells with notes on the role of CD5. Annu Rev Immunol 20:253–300

Bernard A, Coitot S, Bremont A, Bernard G (2005) T and B cell cooperation: a dance of life and death. Transplantation 79:S8–S11

Bikah G, Carey J, Ciallella JR, Tarakhovsky A, Bondada S (1996) CD5-mediated negative regulation of antigen receptor-induced growth signals in B-1 B cells. Science 274:1906–1909

Boes M, Esau C, Fischer MB, Schmidt T, Carroll M, Chen J (1998a) Enhanced B-1 cell development, but impaired IgG antibody responses in mice deficient in secreted IgM. J Immunol 160:4776–4787

Boes M, Prodeus AP, Schmidt T, Carroll MC, Chen J (1998b) A critical role of natural immunoglobulin M in immediate defense against systemic bacterial infection. J Exp Med 188:2381–2386

Bridges CB, Fukuda K, Cox NJ, Singleton JA (2001) Prevention and control of influenza. Recommendations of the Advisory Committee on Immunization Practices (ACIP). MMWR Morb Mortal Wkly Rep 50:1–44

Carvalho TL, Mota-Santos T, Cumano A, Demengeot J, Vieira P (2001) Arrested B lymphopoiesis and persistence of activated B cells in adult interleukin 7(-/)- mice. J Exp Med 194:1141–1150

Chang WLW, Coro ES, Rau FC, Xiao Y, Erle DJ, Baumgarth N (2007) Influenza virus infection causes global respiratory tract B cell response modulation via innate immune signals. J Immunology 178:1457–1467

Cinamon G, Matloubian M, Lesneski MJ, Xu Y, Low C, Lu T et al (2004) Sphingosine 1-phosphate receptor 1 promotes B cell localization in the splenic marginal zone. Nat Immunol 5:713–720

Coro ES, Chang WL, Baumgarth N (2006) Type I IFN receptor signals directly stimulate local B cells early following influenza virus infection. J Immunol 176:4343–4351

Dammers PM, Kroese FG (2005) Recruitment and selection of marginal zone B cells is independent of exogenous antigens. Eur J Immunol 35:2089–2099

Doherty PC (2000) Accessing complexity: the dynamics of virus-specific T cell responses. Annu Rev Immunol 18:561–592

Doherty PC, Topham DJ, Tripp RA, Cardin RD, Brooks JW, Stevenson PG (1997) Effector CD4+ and CD8+ T-cell mechanisms in the control of respiratory virus infections. Immunol Rev 159:105–117

Ferguson AR, Youd ME, Corley RB (2004) Marginal zone B cells transport and deposit IgM-containing immune complexes onto follicular dendritic cells. Int Immunol 16:1411–1422

Fink K, Lang KS, Manjarrez-Orduno N, Junt T, Senn BM, Holdener M et al (2006) Early type I interferon-mediated signals on B cells specifically enhance antiviral humoral responses. Eur J Immunol 36:2094–2105

Flynn KJ, Belz GT, Altman JD, Ahmed R, Woodland DL, Doherty PC (1998) Virus-specific CD8+ T cells in primary and secondary influenza pneumonia. Immunity 8:683–691

Forster I, Rajewsky K (1990) The bulk of the peripheral B-cell pool in mice is stable and not rapidly renewed from the bone marrow. Proc Natl Acad Sci U S A 87:4781–4784

Fujisawa H, Tsuru S, Rtaniguchi M, Zinnaka Y, Nomoto K (1987) Protective mechanisms against pulmonary infection with influenza virus. I. Relative contribution of polymorphonuclear leukocytes and of alveolar macrophages to protection during the early phase of intranasal infection. J Gen Virol 68:425–432

Garcia-Sastre A, Durbin RK, Zheng H, Palese P, Gertner R, Levy DE, Durbin JE (1998) The role of interferon in influenza virus tissue tropism. J Virol 72:8550–8558

Gerhard W, Mozdzanowska K, Furchner M, Washko G, Maiese K (1997) Role of the B-cell response in recover o mice from primary influenza virus infection. Immunol Rev 159:95–103

Godin IE, Garcia-Porrero JA, Coutinho A, Dieterlen-Lievre F, Marcos MA (1993) Para-aortic splanchnopleura from early mouse embryos contains B1a cell progenitors. Nature 364:67–70

Goodnow CC, Cyster JG, Hartley SB, Bell SE, Cooke MP, Healy JI et al (1995) Self-tolerance checkpoints in B lymphocyte development. Adv Immunol 59:279–368

Ha SA, Tsuji M, Suzuki K, Meek B, Yasuda N, Kaisho T, Fagarasan S (2006) Regulation of B1 cell migration by signals through Toll-like receptors. J Exp Med 203:2541–2550

Haas KM, Poe JC, Steeber DA, Tedder TF (2005) B-1a and B-1b cells exhibit distinct developmental requirements and have unique functional roles in innate and adaptive immunity to S. pneumoniae. Immunity 23:7–18

Hao Z, Rajewsky K (2001) Homeostasis of peripheral B cells in the absence of B cell influx from the bone marrow. J Exp Med 194:1151–1164

Hardy RR, Hayakawa K (1991) A developmental switch in B lymphopoiesis. Proc Natl Acad Sci U S A 88:11550–11554

Hardy RR, Hayakawa K (2001) B cell development pathways. Annu Rev Immunol 19:595–621

Hayakawa K, Hardy RR, Herzenberg LA, Herzenberg LA (1985) Progenitors for Ly-1 B cells are distinct from progenitors for other B cells. J Exp Med 161:1554–1568

Heer AK, Shamshiev A, Donda A, Uematsu S, Akira S, Kopf M, Marsland BJ (2007) TLR signaling fine-tunes anti-influenza B cell responses without regulating effector T cell responses. J Immunol 178:2182–2191

Heltemes-Harris L, Liu X, Manser T (2005) An antibody VH gene that promotes marginal zone B cell development and heavy chain allelic inclusion. Int Immunol 17:1447–1461

Herzenberg LA, Stall AM, Lalor PA, Sidman C, Moore WA, Parks DR (1986) The Ly-1 B cell lineage. Immunol Rev 93:81–102

Herzenberg LA, Kantor AB, Herzenberg LA (1992) Layered evolution in the immune system. A model for the ontogeny and development of multiple lymphocyte lineages. Ann N Y Acad Sci 651:1–9

Herzenberg LA, Tung JW (2006) B cell lineages: documented at last! Nat Immunol 7:225–226

Hippen KL, Tze LE, Behrens TW (2000) CD5 maintains tolerance in anergic B cells. J Exp Med 191:883–890

Hogan RJ, Usherwood EJ, Zhong W, Roberts AA, Dutton RW, Harmsen AG, Woodland DL (2001) Activated antigen-specific CD8+ T cells persist in the lungs following recovery from respiratory virus infections. J Immunol 166:1813–1822

Kaminski DA, Stavnezer J (2006) Enhanced IgA class switching in marginal zone and B1 B cells relative to follicular/B2 B cells. J Immunol 177:6025–6029

Kantor AB, Herzenberg LA (1993) Origin of murine B cell lineages. Annu Rev Immunol 11:501–538

Kavaler J, Caton AJ, Staudt LM, Gerhard W (1991) A B cell population that dominates the primary response to influenza virus hemagglutinin does not participate in the memory response. Eur J Immunol 21:2687–2695

Kantor AB, Stall AM, Adams S, Watanabe K, Herzenberg LA (1995) De novo development and self-replenishment of B cells. Int Immunol 7:55–68

Kavaler J, Caton AJ, Staudt LM, Schwartz D, Gerhard W (1990) A set of closely related antibodies dominates the primary antibody response to the antigenic site CB of the A/PR/8/34 influenza virus hemagglutinin. J Immunol 145:2312–2321

Kawikova I, Paliwal V, Szczepanik M, Itakura A, Fukui M, Campos RA et al (2004) Airway hyper-reactivity mediated by B-1 cell immunoglobulin M antibody generating complement

C5a at 1 day post-immunization in a murine hapten model of non-atopic asthma. Immunology 113:234–245

Kenny JJ, Yaffe LJ, Ahmed A, Metcalf ES (1983) Contribution of Lyb 5+ and Lyb 5– B cells to the primary and secondary phosphocholine-specific antibody response. J Immunol 130:2574–2579

Kikuchi K, Kondo M (2006) Developmental switch of mouse hematopoietic stem cells from fetal to adult type occurs in bone marrow after birth. Proc Natl Acad Sci U S A 103:17852–17857

Knoops L, Louahed J, Renauld JC (2004) IL-9-induced expansion of B-1b cells restores numbers but not function of B-1 lymphocytes in xid mice. J Immunol 172:6101–6106

Kraus M, Alimzhanov MB, Rajewsky N, Rajewsky K (2004) Survival of resting mature B lymphocytes depends on BCR signaling via the Igalpha/beta heterodimer. Cell 117:787–800

Kroese FG, Butcher EC, Stall AM, Lalor PA, Adams S, Herzenberg LA (1989) Many of the IgA producing plasma cells in murine gut are derived from self-replenishing precursors in the peritoneal cavity. Int Immunol 1:75–84

Lalor PA, Herzenberg LA, Adams S, Stall AM (1989a) Feedback regulation of murine Ly-1 B cell development. Eur J Immunol 19:507–513

Lalor PA, Stall AM, Adams S, Herzenberg LA (1989b) Permanent alteration of the murine Ly-1 B repertoire due to selective depletion of Ly-1 B cells in neonatal animals. Eur J Immunol 19:501–506

Le Bon A, Thompson C, Kamphuis E, Durand V, Rossmann C, Kalinke U, Tough DF (2006) Cutting edge: enhancement of antibody responses through direct stimulation of B and T cells by type I IFN. J Immunol 176:2074–2078

Lewis GK, Goodman JW, Ranken R (1978) Activation of B cell subsets by T-dependent and T-independent antigens. Adv Exp Med Biol 98:339–356

Lopes-Carvalho T, Foote J, Kearney JF (2005) Marginal zone B cells in lymphocyte activation and regulation. Curr Opin Immunol 17:244–250

Martin F, Oliver AM, Kearney JF (2001) Marginal zone and B1 B cells unite in the early response against T-independent blood-borne particulate antigens. Immunity 14:617–629

Meffre E, Casellas R, Nussenzweig MC (2000) Antibody regulation of B cell development. Nat Immunol 1:379–385

Montecino-Rodriguez E, Leathers H, Dorshkind K (2006) Identification of a B-1 B cell-specified progenitor. Nat Immunol 7:293–301

Mozdzanowska K, Furchner M, Maiese K, Gerhard W (1997) CD4+ T cells are ineffective in clearing a pulmonary infection with influenza type A virus in the absence of B cells. Virology 239:217–225

Mozdzanowska K, Maiese K, Gerhard W (2000) Th cell-deficient mice control influenza virus infection more effectively than Th- and B cell-deficient mice: evidence for a Th-independent contribution by B cells to virus clearance. J Immunol 164:2635–2643

Murakami M, Tsubata T, Shinkura R, Nisitani S, Okamoto M, Yoshioka H et al (1994) Oral administration of lipopolysaccharides activates B-1 cells in the peritoneal cavity and lamina propria of the gut and induces autoimmune symptoms in an autoantibody transgenic mouse. J Exp Med 180:111–121

Noelle RJ (1996) CD40 and its ligand in host defense. Immunity 4:415–419

Ochsenbein AF, Fehr T, Lutz C, Suter M, Brombacher F, Hengartner H, Zinkernagel RM (1999) Control of early viral and bacterial distribution and disease by natural antibodies. Science 286:2156–2159

Pasare C, Medzhitov R (2005) Control of B-cell responses by Toll-like receptors. Nature 438:364–368

Paus D, Phan TG, Chan TD, Gardam S, Basten A, Brink R (2006) Antigen recognition strength regulates the choice between extrafollicular plasma cell and germinal center B cell differentiation. J Exp Med 203:1081–1091

Pecquet SS, Ehrat C, Ernst PB (1992a) Enhancement of mucosal antibody responses to *Salmonella typhimurium* and the microbial hapten phosphorylcholine in mice with X-linked immunodeficiency by B-cell precursors from the peritoneal cavity. Infect Immun 60:503–509

Pecquet SS, Zazulak J, Simpson SD, Ernst PB (1992b) Reconstitution of xid mice with donor cells enriched for CD5+ B cells restores contrasuppression. Ann N Y Acad Sci 651:173–175

Pichlmair A, Schulz O, Tan CP, Naslund TI, Liljestrom P, Weber F, Reis e Sousa C (2006) RIG-I-mediated antiviral responses to single-stranded RNA bearing 5′-phosphates. Science 314:997–1001

Pillai S, Cariappa A, Moran ST (2005) Marginal zone B cells. Annu Rev Immunol 23:161–196

Reading PC, Morey LS, Crouch EC, Anders EM (1997) Collectin-mediated antiviral host defense of the lung: evidence from influenza virus infection of mice. J Virol 71:8204–8212

Renegar KB, Small PA Jr (1991a) Immunoglobulin A mediation of murine nasal anti-influenza virus immunity. J Virol 65:2146–2148

Renegar KB, Small PA Jr (1991b) Passive transfer of local immunity to influenza virus infection by IgA antibody. J Immunol 146:1972–1978

Rolink AG, Schaniel C, Andersson J, Melchers F (2001) Selection events operating at various stages in B cell development. Curr Opin Immunol 13:202–207

Ruprecht CR, Lanzavecchia A (2006) Toll-like receptor stimulation as a third signal required for activation of human naive B cells. Eur J Immunol 36:810–816

Schmitz N, Kurrer M, Bachmann MF, Kopf M (2005) Interleukin-1 is responsible for acute lung immunopathology but increases survival of respiratory influenza virus infection. J Virol 79:6441–6448

Sha Z, Compans RW (2000) Induction of CD4+ T-cell-independent immunoglobulin responses by inactivated influenza virus. J Virol 74:4999–5005

Song H, Cerny J (2003) Functional heterogeneity of marginal zone B cells revealed by their ability to generate both early antibody-forming cells and germinal centers with hypermutation and memory in response to a T-dependent antigen. J Exp Med 198:1923–1935

Stall AM, Adams S, Herzenberg LA, Kantor AB (1992) Characteristics and development of the murine B-1b (Ly-1 B sister) cell population. Ann N Y Acad Sci 651:33–43

Szczepanik M, Akahira-Azuma M, Bryniarski K, Tsuji RF, Kawikova I, Ptak W et al (2003) B-1 B cells mediate required early T cell recruitment to elicit protein-induced delayed-type hypersensitivity. J Immunol 171:6225–6235

Talon J, Horvath CM, Polley R, Basler CF, Muster T, Palese P, Garcia-Sastre A (2000) Activation of interferon regulatory factor 3 is inhibited by the influenza A virus NS1 protein. J Virol 74:7989–7996

Tarlinton DM, McLean M, Nossal GJ (1995) B1 and B2 cells differ in their potential to switch immunoglobulin isotype. Eur J Immunol 25:3388–3393

Virelizier JL, Postlethwaite R, Schild GC, Allison AC (1974) Antibody responses to antigenic determinants of influenza virus hemagglutinin. I. Thymus dependence of antibody formation and thymus independence of immunological memory. J Exp Med 140:1559–1570

Wang X, Li M, Zheng H, Muster T, Palese P, Beg AA, Garcia-Sastre A (2000) Influenza A virus NS1 protein prevents activation of NF-kappaB and induction of alpha/beta interferon. J Virol 74:11566–11573

Wells SM, Kantor AB, Stall AM (1994) CD43 (S7) expression identifies peripheral B cell subsets. J Immunol 153:5503–5515

Yang Y, Tung JW, Ghosn EE, Herzenberg LA, Herzenberg LA (2007) Division and differentiation of natural antibody-producing cells in mouse spleen. Proc Natl Acad Sci U S A 104:4542–4546

The Important and Diverse Roles of Antibodies in the Host Response to *Borrelia* Infections

T. J. LaRocca and J. L. Benach(✉)

1	Introduction	64
2	The Role of Complement.	66
	2.1 Complement-Dependent Antibody-Mediated Killing of *Borrelia*	66
	2.2 Complement Evasion by *Borrelia*.	68
3	The Role of B Cells and Antibodies in the Host Response to *Borrelia*.	69
	3.1 T Cell-Dependent B Cell Responses in *Borrelia* Infection.	69
	3.2 T Cell-Independent B Cell Responses in *Borrelia* Infection	72
	3.3 The Protective Role of Antibodies in Experimental Models	74
	3.4 Antibody Responses and the Clinical Setting.	77
	3.5 Antibody Responses Associated with Specific Tissue Manifestations of *Borrelia* Infection.	79
	3.6 Antibody-Mediated Autoimmunity in *Borrelia* Infection	80
4	Unique Properties of Antibodies in *Borrelia* Infection	82
	4.1 Immune Pressure and Development of Escape Mutants.	82
	4.2 Antibody Interactions Within the Tick Vector	82
	4.3 The Presence and Importance of Complement-Independent, Bactericidal Antibodies.	84
References		88

Abstract Antibodies are of critical importance in the host response to tick-borne *Borrelia* species that cause relapsing fever and Lyme disease. Recent studies on the role of various B cell subsets in the host response to *Borrelia*, complement-independent, bactericidal antibodies, and diagnostics led to this review that focuses on the array of functions that antibodies to *Borrelia* can perform.

Abbreviations Arp: Arthritis-related protein; C4BP: C4b-binding protein; CNS: Central nervous system; CRASP: Complement regulator acquiring surface protein; CSF: Cerebrospinal fluid; Dbp: Decorin-binding protein; DHS: Downstream homology sequence; d.p.i.: Days postinoculation; ELISA: Enzyme-linked

---

J. L. Benach
Center for Infectious Diseases, 5120 Centers for Molecular Medicine, Stony Brook,
NY 11794-5120, USA
e-mail: jbenach@notes.cc.sunysb.edu

immunosorbent assay; Fab: Fragment antigen binding; Fc: Fragment crystallizable; FcRn: Neonatal Fc receptor; Fhb: Factor H-binding protein; FO B cells: Follicular B cells; GlpQ: Glycerophosphodiester phosphodiesterase; HUVEC: Human umbilical vein endothelial cells; IFA: Indirect immunofluorescence assay; LPS: Lipopolysaccharide; MAb: Monoclonal antibody; MAC: Membrane attack complex; MZ B cells: Marginal zone B cells; OM: Outer membrane Osp: Outer surface protein; scFv: Single-chain variable fragment; SCID: Severe combined immunodeficiency; TRLA: Treatment-resistant Lyme arthritis; UHS: Upstream homology sequence; Vlp: Variable large protein; VlsE: Vmp-like sequence, expressed; Vmp: Variable major protein; Vsp: Variable small protein; w.p.i.: Weeks postinoculation

1 Introduction

Lyme disease and relapsing fever are arthropod-borne, spirochetal infections that can affect the skin, joints, heart and nervous system. *Borrelia burgdorferi* was first identified as the causative agent of Lyme disease (Benach et al. 1983; Burgdorfer et al. 1982), but this disease can also be caused by several other genospecies (*B. garinii*, *B. afzelii*, *B. japonica*), and there are others (*B. andersoni*, *B. lusitaniae*, *B. valaisiana*) whose role in human disease have not been directly proven (Anderson et al. 1988, 1989; Baranton et al. 1992; Canica et al. 1993; Kawabata et al. 1993; Le Fleche et al. 1997; Wang et al. 1997). Relapsing fever is also caused by various *Borrelia* species (spp.) that can be found across the world, and these include the tick-transmitted *B. hermsii*, *B. turicatae*, *B. parkeri*, *B. mazzotti*, *B. venezuelensis*, *B. duttonii*, *B. crocidurae*, *B. persica*, *B. hispanica*, *B. latyschewii*, *B. caucasia*, and *B. recurrentis*, the last of which is transmitted by lice (reviewed in Connolly and Benach 2005; Cutler et al. 1997; Ras et al. 1996). Hard-bodied ticks of the *Ixodes* genus are the primary vectors of Lyme disease *Borrelia*, whereas relapsing fever *Borrelia* are commonly transmitted by soft-bodied ticks of the genus *Ornithodoros* (Burgdorfer et al. 1982; Burgdorfer and Gage 1986; Burroughs and Holdenried 1944; Davis 1939, 1940; Piesman and Sinsky 1988).

Acute manifestations of both diseases include fatigue, fever, headache, arthralgia, or myalgia. Lyme disease differs in that it is often marked by a reddish, bullseye-like skin lesion at the site of the tick bite called erythema migrans. In relapsing fever, acute manifestations evolve more rapidly and tend to be more severe, sometimes resulting in death, as a result of the high density bacteremia (spirochetemia) caused by these spirochetes (see Fig. 1) (Southern and Sanford 1969). The spirochetemia in relapsing fever is characterized by an initial large peak that occurs after spirochetes disseminate from dermal tissue to the blood. This peak is rapidly cleared by antibodies and it is followed by a series of smaller peaks that develop as a result of antigenic variation of these spirochetes (Barbour et al. 1982; Coffey and Eveland 1967b; Stoenner et al. 1982). The antigenic variation is such that each peak is

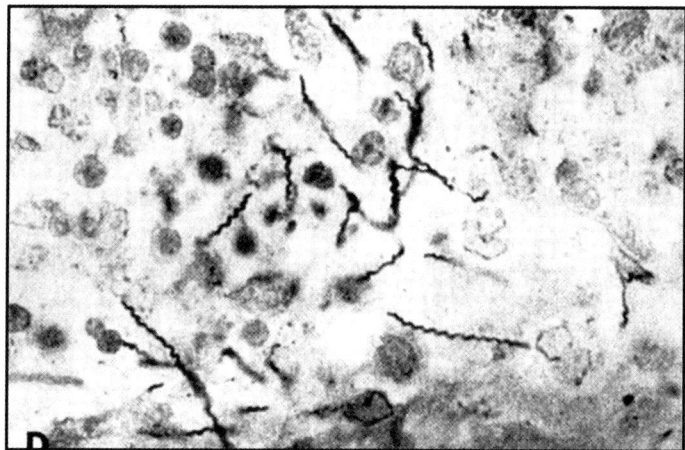

Fig. 1 Relapsing fever infection. Shown is an image from a 5 mm paraffin thin section of murine liver during peak spirochetemia with silver-stained *Borrelia*. The presence of a great number of spirochetes in the liver sinusoids demonstrates the impressive high density spirochetemia that is characteristic of relapsing fever *Borrelia*

predominantly composed of spirochetes of one serotype, with different serotypes present at much lower densities. As the immune system develops a response to the major serotype, spirochetes of that serotype are cleared, allowing for organisms expressing another serotype to thrive and causing a subsequent peak (Barbour et al. 1982; reviewed in Johnson 1977; Stoenner et al. 1982). The antigenic switch occurs at the genetic level via transfer of silent variable major protein (*vmp*) genes that become expressed one at a time and are interchangeable (reviewed in Barbour 1990; Burman et al. 1990; Meier et al. 1985; Plasterk et al. 1985). Once translated, the expressed variable major protein (Vmp) antigen becomes the dominant surface protein of the spirochete, producing a serotype. The insertion of a previously silent (archival) *vmp* gene into the expression plasmid depends on recombination between upstream and downstream homology sequences (UHS and DHS, respectively) that flank the genes (Dai et al. 2006). Frequency of antigenic switch depends upon identity between UHS sequences and placement of DHS sequences on different *vmp* genes, indicating a programmed method of antigenic variation (Barbour et al. 2006). Due to this programmed antigenic variation, there is a hierarchy of switching that occurs, with certain Vmps favoring expression in early infection and others favoring expression in later infection. This allows for the antigenic variation to be semi-predictable, and so, even though there may be many antigenically variable serotypes, the relapse cycle is self-limited (Barbour et al. 2006).

The *Borrelia* spirochetes have the morphology of a Gram-negative bacterium, with an outer membrane (OM), periplasm, and a thin cell wall. However, spirochetes do not have a lipopolysaccharide (LPS) that contains lipid A, and the periplasmic flagella, which serve structural and motile roles, appear to be unique to these bacteria (reviewed in Barbour and Hayes 1986; Beck et al. 1985; reviewed in Johnson 1977; Takayama et al. 1987). Major antigens of *Borrelia* are the surface-exposed lipoproteins present in the OM (Brandt et al. 1990). In *B. burgdorferi* these are referred to as outer surface proteins (Osps) (reviewed in Barbour 1991; Benach et al. 1988; Bergstrom et al. 1989; Carter et al. 1994; Fraser et al. 1997; Howe et al. 1985). The most prominent antigens of relapsing fever *Borrelia* are collectively referred to as Vmps but can be divided into two groups of lipoproteins based on their molecular weight: the variable small proteins (Vsps) of 19–22 kDa and variable large proteins (Vlps) of 35–40 kDa (reviewed in Barbour 1991). As *Borrelia* are extracellular pathogens, humoral immunity is an important method of defense. Antibodies are, therefore, important weapons in combating these infections. In this review, we will emphasize the more unique aspects of antibodies to *Borrelia*, and their relationship to the other components of the humoral branch of the immune response. Complement and antibodies act in concert to eliminate pathogenic bacteria, yet in *Borrelia* infections this synergy is not so apparent.

2 The Role of Complement

The complement system (see Fig. 2) makes up one arm of humoral immunity with antibodies representing the other and both are important in the defense against extracellular pathogens. The classical, alternative, and mannan-binding lectin complement pathways differ in their initial steps and method of recruitment but produce similar results that include opsonization, inflammatory cell recruitment, and the formation of the membrane attack complex (MAC–lytic complement) (Janeway et al. 2001).

2.1 Complement-Dependent Antibody-Mediated Killing of Borrelia

Complement-dependent antibody-mediated killing of *B. burgdorferi* via the classical pathway has been documented in vitro (Kochi and Johnson 1988). This complement-dependent killing was rather unusual because fragment antigen-binding (Fab) fragments were able to mediate complement killing. As the fragment crystallizable (Fc) is the complement-activating domain, this complement-dependent effect was presumably the result of Fab binding, which in turn altered steric conformation to allow for proper MAC deposition (Kochi et al. 1991, 1993). Many monoclonal antibodies (MAbs) and antisera to *B. burgdorferi* have also been shown to be dependent on complement for bactericidal activity in vitro (Aydintug et al. 1994; Lovrich et al. 1991; Ma et al. 1995; Munson et al. 2000, 2006; Nowling and Philipp 1999; Remington et al. 2001; Rousselle et al. 1998; Sole et al. 1998). These numerous in vitro accounts underscore a role for antibody-mediated killing dependent

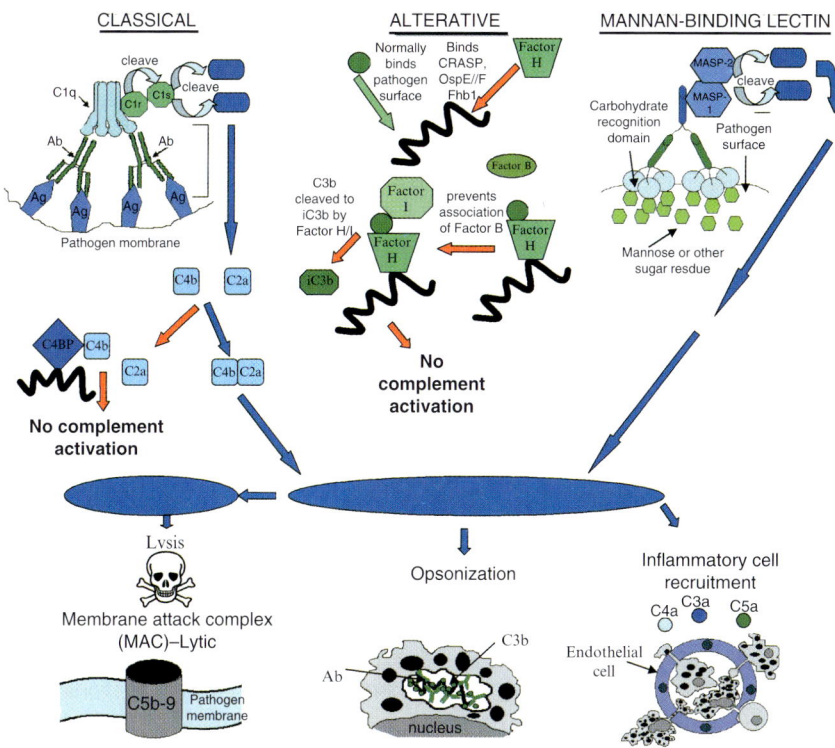

Fig. 2 The complement system. Complement deposition can occur through three pathways, the classical, alternative, and mannan-binding (MB) lectin pathways, which differ in their initial steps and method of recruitment. In the classical pathway, the C1 complex (C1q with bound C1r and C1s) binds the antibodies of already formed Ab:Ag immune complexes. After being cleaved by C1r, C1s cleaves C2 and C4 to C2a and C4b. Components C2a and C4b can associate to form C3 convertase allowing for complement activation. *B. recurrentis* and *B. duttonii* have been shown to bind complement regulator C4b binding protein (C4BP) which inhibits association of C2a and C4b. C4b is subsequently cleaved by Factor I (not shown), and complement activation is prevented. It is possible that relapsing fever *Borrelia* evade the classical pathway in this way. In the alternative pathway, a series of hydrolysis and cleavage events leads to the production of C3b which can bind covalently to the pathogen membrane. Factor B normally binds C3b forming a C3 convertase that allows for complement activation. However, *Borrelia* bind complement regulator Factor H via CRASPs, OspE/F, or Fhb1 which inhibits the association of C3b and Factor B. Factor H catalyzes the cleavage of C3b by Factor I and complement activation is prevented. In the MB lectin pathway, carbohydrate recognition domains of the mannan-binding lectin bind appropriately spaced sugar residues, such as mannose, on the pathogen surface. If complement activation occurs (C3 and C5 convertases formed) complement effector functions, common to all 3 pathways, can occur. The effector functions include the following: 1. recruitment of inflammatory cells 2. Opsonization (coating) of pathogens for enhanced removal by phagocytes. 3. Formation of the membrane attack complex (MAC). The MAC forms a lytic pore in the pathogen membrane and has traditionally been thought of as the sole method whereby antibodies can exert a bactericidal effect in the absence of phagocytic cells

on complement in immunity against *Borrelia*, as would be expected in the host response to an extracellular bacterium.

2.2 Complement Evasion by Borrelia

Despite the evidence described in the previous section, recent studies have shown that *Borrelia* develops resistance to certain forms of complement-dependent killing, particularly by evasion of the alternative complement pathway (see Fig. 2) (reviewed in Kraiczy et al. 2001c; van Dam et al. 1997). In Lyme disease, this evasion is mediated by the binding of the complement regulator-acquiring surface proteins (CRASPs) and proteins of the OspE/F Erp family, present on the spirochete surface, to host Factor H and Factor H-like protein-1 (FHL-1) (Alitalo et al. 2002; Brooks et al. 2005; Hartmann et al. 2006; Hellwage et al. 2001; Hovis et al. 2006c;2001b, 2003, Kraiczy et al. 2004; Rossmann et al. 2006). Complement evasion is made possible by Factor H and FHL-1, which are host alternative complement pathway regulators, acting as cofactors for Factor I. This in turn promotes the degradation of C3b and C3 convertase and thus inhibits the lytic effects of complement. The CRASPs bind to the C-terminus of Factor H or FHL-1, orienting them in such a way as to allow for the continuation of their regulatory function, contributing to complement evasion by *Borrelia* (Hartmann et al. 2006; Kraiczy et al. 2001a, 2004). This binding interaction is quite intricate because it seems to depend on the existence of a discontinuous binding site, in which charge–charge and hydrophobic interactions both contribute, and is stabilized by C-terminal CRASP residues (Cordes et al. 2005, 2006). There has been some debate as to whether CRASP-1 has a significant role in complement evasion, as it was shown that patient serum did not react against CRASP-1 in immunoblots (McDowell et al. 2006). Moreover, upregulation of the CRASP-1 gene did not occur in *B. burgdorferi* recovered from infected mice (McDowell et al. 2006). However, patient serum has shown reactivity to this protein in its native form, suggesting that it is expressed in human infection and relevant to complement evasion in vivo (Rossmann et al. 2006; von Lackum et al. 2005). Proteins of the OspE/F Erp family also function in a similar manner, although their binding may depend on the presence of coiled-coil domains, with contributing C-terminal lysine residues (Alitalo et al. 2004; Hellwage et al. 2001; Kraiczy et al. 2003; McDowell et al. 2004). These complement evasion proteins have been shown to possess the ability to bind Factor H from different animal hosts, presumably contributing to the host range of Lyme disease *Borrelia* (Hovis et al. 2006c; Stevenson et al. 2002). Similar alternative complement evasion occurs in relapsing fever infection as well (Hovis et al. 2004; McDowell et al. 2003). This is mediated by Factor H-binding protein A1 (FhbA1) and FhbA2, which are functionally similar to the CRASP and OspE/F proteins. Their binding is contingent on N- and C-terminal residues and may depend on loop or coiled-coil domains (Hovis et al. 2006a, 2006b). Since *Borrelia* has been shown to be resistant to the effects of complement via the alternative pathway, the other major constituents of humoral immunity, antibodies, certainly acquire greater importance in the host response.

Interestingly, relapsing fever spirochetes *B. recurrentis* and *B. duttonii* were shown to bind C4b-binding protein (C4BP), the functional homolog to Factor H in the classical complement pathway (see Fig. 2) (Meri et al. 2006). This binding was achieved in a manner similar to the binding of Factor H by *Borrelia*. This suggests that relapsing fever *Borrelia* may be capable of evading destruction via the classical complement pathway as well. If this is the case, evasion of complement deposition may contribute to the survival of relapse populations in the infection, and more importantly, may explain the importance of complement-independent, bactericidal antibodies associated with immunity to relapsing fever (see Sect. 3.3) (Barbour and Bundoc 2001; Connolly and Benach 2001; reviewed in Connolly and Benach 2005; Connolly et al. 2004; Newman and Johnson 1981).

3 The Role of B Cells and Antibodies in the Host Response to *Borrelia*

B cells can be divided into two main sets, B1 and B2, each with two subsets (see Fig. 3) (Janeway et al. 2001; Martin and Kearney 2001). B1 B cells, which contain the B1a and B1b subsets, are T cell-independent, self-renewing B cells found mainly in the pleural and peritoneal cavities (Janeway et al. 2001; Martin and Kearney 2001; see also the chapter by N. Baumgarth et al., this volume). These B cells are responsible for natural IgM (Ochsenbein et al. 1999), but can be driven to expand and secrete specific IgM, which is important in the early responses to some infections (Alugupalli et al. 2003, 2004; Baumgarth et al. 2000; Haas et al. 2005; Montecino-Rodriguez and Dorshkind 2006; see also the chapter by K.R. Alugupalli, this volume). Mature B2 B cells, which contain follicular (FO) and marginal zone (MZ) B cell subsets, account for the majority of B cells found within a host. Follicular B cells are the textbook examples of B cells, which require T cell help, generally of the T helper-2 (T_{H2}) phenotype (Janeway et al. 2001; Rohrer et al. 1983). Marginal zone B cells are T cell-independent, IgM-secreting B cells found in the MZ at the border of the white pulp in the spleen and do not recirculate (Guinamard et al. 2000; Janeway et al. 2001; Kumararatne et al. 1981; Kumararatne and MacLennan 1981, 1982; Oliver et al. 1999; Pillai et al. 2005). In infection with Lyme disease and relapsing fever *Borrelia*, there are both T cell-dependent and T cell-independent antibody responses. This review will consider both mechanisms and the contexts in which they operate.

3.1 T Cell-Dependent B Cell Responses in Borrelia Infection

In Lyme disease, FO B cells are considered to be of particular importance in immunity, as a T_{H2} response is associated with resolution of infection and disease (see Fig. 3) (Kang et al. 1997; Matyniak and Reiner 1995; Rohrer et al. 1983). This is apparently based upon the cytokines elicited by *B. burgdorferi* infection

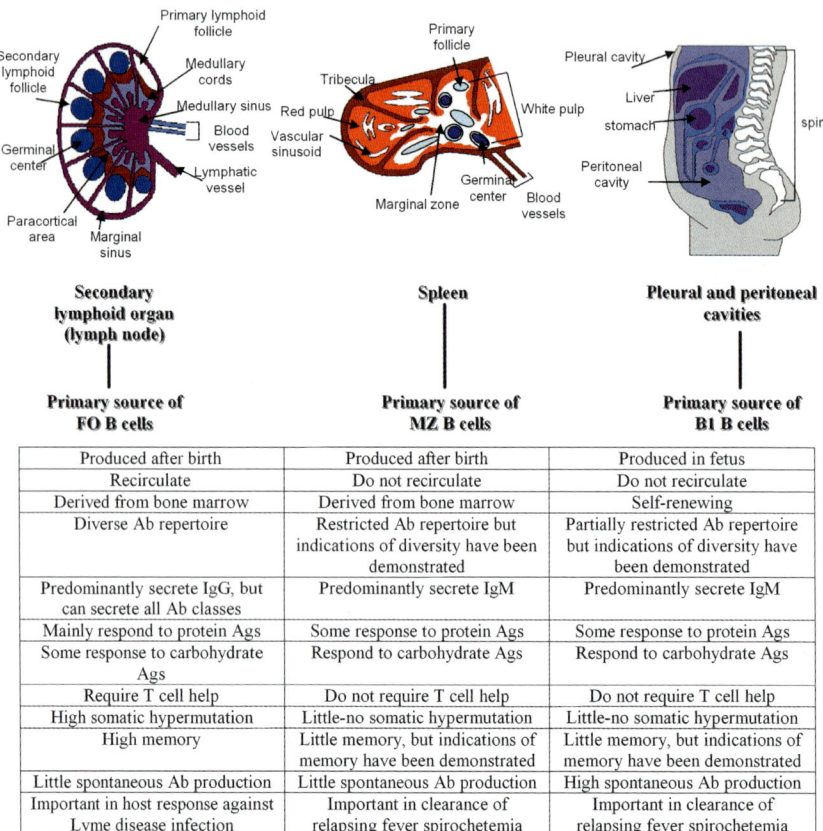

Fig. 3 Different B cell subsets and their locations. Follicular (FO) B cells are commonly located in secondary lymphoid organs such as the spleen and lymph nodes (depicted). FO B cells can be found within the primary and secondary lymphoid follicles in lymph nodes. Within the secondary lymphoid follicles are regions of intense B cell proliferation called germinal centers in which T cell dependent B cell responses occur (class switching, somatic hypermutation, affinity maturation). FO B cells elicited by a T_{H2} response are known to be important in the host response against Lyme disease infection (*B. burgdorferi*), however there is evidence that this immunity occurs in the absence of T cells, as well. Marginal zone (MZ) B cells are located in the marginal zone at the border of the white pulp in the spleen as indicated in the diagram. B1 B cells are commonly found in body cavities such as the pleural and peritoneal cavities depicted in the diagram. Both MZ and B1b B cells have been shown to be important in the host response to relapsing fever *Borrelia* (*B. hermsii*) and function in the absence of T cell help. These cells produce serotype-specific IgM which is critical for the clearance of relapsing fever spirochetemia. Notable characteristics of each B cell subset are listed in the table

in BALB/c and C3H/HeN (or C3H/HeJ) mice (Matyniak and Reiner 1995). In C3H mice, where a severe, nonresolving arthritis phenotype is always seen, IFN-γ is the predominant cytokine elicited. This is consistent with a T_{H1}-type cellular immunity, in which inflammation would play a large part and would be a contributor to Lyme arthritis, whereas neutralization of IFN-γ attenuates disease (Isakson et al. 1982; Matyniak and Reiner 1995; Stevens et al. 1988). In contrast, BALB/c mice, which are associated with much milder disease that eventually resolves, have a predominant IL-4 response to *B. burgdorferi*. This response is consistent with a T_{H2}-type humoral immunity, as IL-4 is a cytokine important for inducing antibody class switch in B cells (Matyniak and Reiner 1995; Snapper and Paul 1987; Stevens et al. 1988). At 2 days postinoculation (d.p.i.) in BALB/c mice, the predominant response is T_{H1}, which may provide a first line of defense, followed by the emergence of the T_{H2} response at 14 d.p.i and lasting until 30 d.p.i., at which point arthritis resolves, clearly indicating the importance of FO B cells (Kang et al. 1997; Matyniak and Reiner 1995). Moreover, transfer of a *B. burgdorferi*-reactive T_{H2} cell line to naïve mice conferred protection, and depletion of CD4 T cells in BALB/c and C3H/HeN mice increased arthritis severity as well as spirochete numbers in joints and skin (Keane-Myers and Nickell 1995b; Rao and Frey 1995). Adoptive transfer of T cell-enriched lymphocyte populations from previously infected mice to infected SCID mice led to mild improvement in Lyme arthritis and carditis (Barthold et al. 1996). In C3H mice, there is an increase in IgG2a and decrease in IgG1, characteristic of a T_{H1} response, and in BALB/c mice there is an increase in IgG1, the characteristic antibody subclass elicited in a T_{H2} response (Isakson et al. 1982; Keane-Myers and Nickell 1995a; Snapper and Paul 1987; Stevens et al. 1988). Treatment of C3H mice with rIL-4 early in infection led to greater control of the infection as joint swelling and spirochete numbers decreased (Keane-Myers et al. 1996). Treatment with rIL-4 also led to a decrease in IFN-γ, IgG2a, and IgG3 and an increase in T_{H2}-associated IgG1 in vitro (Keane-Myers et al. 1996). Experiments with IL-4$^{-/-}$ BALB/c mice demonstrated the development of a T_{H1} response, as IgG2a and carditis severity increased, in contrast to the less severe carditis seen in IL-4$^{+/+}$ mice (Satoskar et al. 2000). Further demonstration of the involvement and importance of T_H humoral immunity in response to Lyme disease was seen as active immunization of C3H/HeJ mice with rOspA elicited an OspA T cell epitope, which induced the production of OspA-reactive IgG (Bockenstedt et al. 1996). Also, in vitro co-incubation of T and B cells from hamsters 2 weeks after vaccination with *B. burgdorferi* allowed for the production of bactericidal antibodies (Jensen et al. 1998). However, T cells from 1 or 3 weeks postinoculation (w.p.i.) suppressed the production of bactericidal antibodies, suggesting a dual induction/suppression role for T cells in the antibody response against *B. burgdorferi* (Jensen et al. 1998). A possible factor in promotion of a T_{H2} response is IL-6, as IL-6-deficient mice developed arthritis faster and had reduced IgG2b, consistent with a lowered T_{H2} response (Anguita et al. 1998). The need for T cells in *B. burgdorferi* immunity is further underscored by the induced migration of T cells across human umbilical vein endothelial cells (HUVEC) in response to *B. burgdorferi*, in vitro (Dame et al. 2007; Gergel and Furie 2001, 2004). That the T_{H2} response develops in mouse strains that are more resistant to infection such as BALB/c and C57BL/6, is an indication of the strong protective role of antibodies that are associated with this response.

3.2 T Cell-Independent B Cell Responses in Borrelia Infection

Though T_{H2}-type immunity was shown to be important in helping the B cell antibody response in Lyme disease infection, there is evidence suggesting that T cells are not necessary to battle *B. burgdorferi* infection successfully. Initial evidence was observed in CB-17 SCID mice, as presensitized B cells alone conferred partial protection against *B. burgdorferi* (Schaible et al. 1994). Further evidence was shown in BALB/c, C3H/HeN, and C57BL/6 mice previously immunized with *B. burgdorferi* soluble sonicate antigen and then challenged at different times. Both T_{H1}- and T_{H2}-type antibodies were elicited early in the response after immunization and infection, however, without production of the T_{H2}-associated IL-4 (Frey and Rao 1995). In comparison to controls, there was no difference in arthritis and remission in CD40 ligand (CD40L)-deficient mice, which would be defective in T cell help, as CD40L is a T cell costimulatory molecule responsible for antigen presenting cell activation. Sera from these mice passively protected SCID mice, suggesting that T cell help, in this setting, may be expendable (Fikrig et al. 1996). Sera from class II transactivator (CIITA)-deficient mice (defective in antigen presentation) were also shown to be protective in C3H and SCID mice (Fikrig et al. 1997). T cell-deficient (deficient in all T cells) mice were shown to have no effect on the resolution of Lyme arthritis and carditis, and sera from these mice protected naïve mice, but it did not resolve arthritis when passively administered to SCID mice (McKisic and Barthold 2000). These experiments indicate that T cell-independent antibody responses to *B. burgdorferi* infection can occur.

T cell-independent antibody responses also occur in relapsing fever infection and seem to be the predominant method for the clearance of spirochetemia in this disease (see Fig. 3). Neonatally thymectomized mice, nude mice (deficient in $\alpha\beta$ T cells), and mice deficient in all T cells cleared *B. turicatae* with similar efficiency to normal mice, providing the first piece of evidence that T cells are not needed for an efficient immune response against relapsing fever (Newman and Johnson 1984). B cells were shown to be responsible for immunity, and IgM was the predominant antibody class produced even when T cells were present (Newman and Johnson 1984). Furthermore, nude mice had no defect in clearing *B. hermsii* infection, as well (Barbour and Bundoc 2001). A robust, early IgM response was shown to be responsible for relapsing fever clearance in vivo (Connolly and Benach 2001), and mice incapable of secreting IgM were unable to clear infection, suggesting that IgM is critical and other antibody classes are not, as would be the case in a T cell-independent response (Alugupalli et al. 2003). Additionally, activation-induced deaminase-deficient mice ($AID^{-/-}$) cleared *B. hermsii* without a problem, suggesting a T cell-independent response, as there was no antibody class switch (Alugupalli et al. 2004). Mice of SCID and $Rag^{-/-}$ phenotypes could not clear *B. hermsii* infection, pointing to the importance of lymphocytes and antibodies in relapsing fever clearance. It was further shown that IgM induction by B1b B cells, a subset known to function independently of T cell help, contributed to the resolution of relapsing fever caused by *B. hermsii* (Alugupalli et al. 2003). Mice deficient in T cells ($TCR^{-/-}$) and both

T and FO B cells (IL-7$^{-/-}$) exhibited efficient *B. hermsii* clearance, suggesting the importance of other B cell subsets, such as B1 and MZ B cells. Those mice with a depleted B1 cell population (X-linked immunodeficiency [*xid*] mice) exhibited longer, more severe spirochetemia, and were even driven to expand their already depleted B1b B cells after *B. hermsii* resolution, clearly suggesting an important role for this B cell subset in relapsing fever immunity (Alugupalli et al. 2003). The immunity afforded by this subset of B cells in relapsing fever infection was shown to be long-lasting, as primed B1b B cells from convalescent mice conferred protection to Rag$^{-/-}$ mice upon reconstitution and induced the rapid production of *B. hermsii*-specific IgM (Alugupalli et al. 2004). This certainly suggested a role for B1b B cells in a T cell-independent, IgM memory response. These findings also seemed to implicate B1b B cells as the most important B cell subset in the early host response to *B. hermsii*, as splenectomized mice (deficient in MZ B cells) also cleared infection. However, these mice could not clear spirochetemia when a low-passage, virulent *B. hermsii* strain was used, indicating that B1b B cells are not the only important subset in this host response (Alugupalli et al. 2003). Indeed, another T cell-independent B cell subset, MZ B cells, was shown to be important in relapsing fever immunity, and may work cooperatively with B1b B cells (Belperron et al. 2005). Marginal zone B cells become activated after association with virulent *B. hermsii*, and loss of activation is coincident with IgM production and spirochete burden reduction. Mice depleted of MZ B cells, by anti-LFA-1 and anti-α 4β1 integrin MAb treatment, had reduced anti-*B. hermsii* IgM and increased pathogen burden. CD1d (a nonclassical MHC molecule predominant on MZ B cells and important in the recognition of lipid antigens) was shown to have a role in spirochete reduction and anti-*B. hermsii* antibody production from MZ B cells. CD1d$^{-/-}$ mice had similar results to MZ B cell-depleted mice in terms of spirochete burden and were defective in MZ B cell IgM secretion (Belperron et al. 2005). This may suggest a new function for CD1d, since traditionally CD1d has been known to be involved in lipid antigen presentation to natural killer T cells, which in turn promote IgM and IgG1 production, provide T_H cytokines, and have immune regulatory function (reviewed in Bendelac et al. 2001; reviewed in Galli et al. 2003; reviewed in Vincent et al. 2003; Yang et al. 2003). As MZ B cells in CD1d$^{-/-}$ mice retained an activated state longer than those from wild type mice, the involvement of CD1d in immune regulation of MZ B cells is suggested during the host response to *B. hermsii* (Belperron et al. 2005). It could be that CD1d is providing an autocrine signal to MZ B cells (without T cell assistance) promoting the production of *Borrelia*-specific IgM in response to *B. hermsii* antigens (Belperron et al. 2005). Recently, MyD88, a toll-like receptor adaptor molecule, was shown to contribute to the kinetics of the IgM response against *B. hermsii* (Bolz et al. 2006). MyD88$^{-/-}$ mice displayed a very slight delay in IgM response that was overcome by 6 d.p.i . Despite this delay, antibodies from these mice could clear spirochetemia when transferred to SCID, so the precise role that MyD88 may play in this is still unclear (Bolz et al. 2006). These results collectively give strong support to the idea that antibody-mediated immunity against relapsing fever infection is largely T cell-independent, and that rapid production of IgM by B1b and MZ B cells is a necessity for an efficient host response and immunological memory to these pathogens (see also chapter by K.R. Alugupalli, this volume).

The mitogenic properties of *Borrelia* lipoproteins are also suggestive of T cell-independent responses and may be indicative of what is occurring in the T cell-independent responses against Lyme disease and relapsing fever. *Borrelia burgdorferi* sonicate and membrane blebs stimulated proliferation and differentiation of B cells into antibody-secreting cells (IgM) in the absence of accessory cells and concomitant with IL-6 production in vitro (Schoenfeld et al. 1992; Whitmire and Garon 1993). These results were replicated in vivo, and OspA and OspB were specifically shown to be capable of B cell mitogenic stimulation (Ma and Weis 1993; Yang et al. 1992). OspA and OspB were capable of inducing IgM production in vitro and both proteins were shown to have equal activity, suggesting that this may be a property of all *Borrelia* lipoproteins (Ma and Weis 1993). These effects were also seen in human B cells, as *B. burgdorferi* sonicate or OspA were mitogenic to these cells and stimulated IL-6 in vitro (Tai et al. 1994). Another *B. burgdorferi* antigen consisting of lipoprotein and glycolipid induced proliferation of B cells and antibody secretion in vitro (Honarvar et al. 1994). This indicates the involvement of many different *Borrelia* antigens in B cell mitogenic stimulation. The inherent mitogenic properties of *Borrelia* lipoproteins may drive the development of a T cell-independent antibody response in these infections.

3.3 The Protective Role of Antibodies in Experimental Models

Bactericidal antibodies are of extreme importance in the host response to *Borrelia*, as antisera have been shown to be bactericidal, and a number of bactericidal MAbs, directed against OspA, OspB, and OspC in *B. burgdorferi*, and different Vmps in relapsing fever *Borrelia*, have been discovered and investigated (Aydintug et al. 1994; Barbour and Bundoc 2001; Coleman et al. 1992; Connolly et al. 2004; Ma et al. 1995; Munson et al. 2000; Rousselle et al. 1998; Sadziene et al. 1994). Bactericidal antibodies of IgG1, IgG2a, and IgG2b isotypes against OspA appear to be dependent upon complement for their action and are produced extensively in response to IL-6 in vitro. In addition, IL-6 also causes an expansion in B cell number, seen previously upon mitogenic B cell stimulation (discussed in Sect. 3.2) (Munson et al. 2006). Treatment with IL-4 or IFN-γ inhibit the production of these antibodies in vitro with anti-IL-4 or anti-IFN-γ restoring antibody production (Munson et al. 2000, 2002). This seems to contrast with what was discussed in Sect. 3.1; however, these in vitro cultured B cells have already undergone class switch as they produce IgG. Thus, one may not expect IL-4 to have an effect, while IL-6 may be important for in vitro B cell proliferation and stimulation of antibody secretion. Interestingly, B cells treated with anti-IFN-γ cause severe destructive arthritis in vivo, perhaps suggesting an overwhelming antibody response that may promote pathology (Munson et al. 2004). Exogenous macrophages were required in conjunction with IL-6 to cause increased anti-OspA bactericidal antibody production, suggesting that macrophages may process *Borrelia* antigens and elicit IL-6 for antibody production (Munson et al. 2006). However, macrophages were not required for the IL-6-stimulated production of anti-OspC bactericidal IgG2b in vitro (Remington et al. 2001). These findings indicate some of the important factors required for antibody production/suppression in *Borrelia* infection, particularly those which rely on complement-mediated lysis.

The first demonstration of the vast importance of antibodies in defense against *Borrelia* infection in vivo was the finding that passive transfer of immune sera is protective against these infections. Passive transfer of rabbit immune serum to hamsters conferred sufficient protection against *B. burgdorferi*, and was the first demonstration of the protective capacity of antibodies in this infection (Johnson et al. 1986). Subsequently, antibodies have been shown to be the primary mediators of this protection, as immune serum was shown to protect SCID mice, which are deficient in both B and T cells (Barthold et al. 1996, 1997, 2006; Schaible et al. 1990; Zhong et al. 1997). Indeed, this has been studied extensively, and the protective capacity of sera against specific Ags, and monoclonal antibodies (MAbs) derived from these immune sera, was assessed (Barthold and Bockenstedt 1993; Barthold et al. 1997, 2006; Fikrig et al. 1994; Mbow et al. 1999; Probert et al. 1997; Schaible et al. 1990; Schmitz et al. 1992; Zhong et al. 1997). Anti-OspC serum was shown to be effective therapeutically in eliminating spirochetes. It also induced arthritis and carditis resolution in SCID mice (Zhong et al. 1997, 1999). Anti-OspB serum was effective in growth inhibition of *B. burgdorferi* as well as protection of C3H/HeJ mice (Probert et al. 1997; Zhong et al. 1997, 1999). Passive transfer of MAbs, reactive against arthritis-related protein (Arp) and decorin-binding protein A (DbpA), to SCID mice was also shown to induce arthritis and carditis remission (Barthold et al. 2006). Only anti-DbpA was protective, however, and neither sera caused reduction in tissue spirochetes, when administered after infection, pointing to the importance of antibodies reactive to other antigens (Barthold et al. 2006). Anti-OspA IgG2a and IgG2b conferred protection in SCID mice. Immunoglobulin G2 from 3-week immune serum protected hamsters, and anti-OspC IgG2a protected C3H/HeJ mice, underscoring the importance of this antibody class in protection against *B. burgdorferi* (Mbow et al. 1999; Schaible et al. 1990; Schmitz et al. 1992).

The importance of antibodies in *Borrelia* immunity can also be seen in the success of active immunization studies specific for *B. burgdorferi* infection. These studies made use of OspA (by itself or delivered on bacterial or viral surfaces, delivered orally or by syringe), truncated OspA, OspC, Omp66, and DbpA, all of which elicited protective antibodies as the main mediators of protection (Bockenstedt et al. 1997; Dunne et al. 1995; Exner et al. 2000; Fikrig et al. 1990, 1991, 1992a; Koide et al. 2005; Luke et al. 1997; Nassal et al. 2005; Scheckelhoff et al. 2006; Simon et al. 1991; Skamel et al. 2006; Ulbrandt et al. 2001).

Antibodies are also of critical importance to immunity against the various tick- and louse-borne relapsing fever species of *Borrelia* and play a critical role in clearance of these pathogens from the blood. The protective capacity of passively transferred immune sera was demonstrated several times in this infection, the first demonstration dating as far back as 80 years prior to this same discovery in Lyme disease (Calabi 1959; Novy 1906; Stoenner et al. 1982). Subsequently, serotype-specific IgM antibodies were shown to mediate passive protection of BALB/c and SCID mice against infection with different relapsing fever *Borrelia* spp. (Barbour and Bundoc 2001; Connolly et al. 2004; Yokota et al. 1997). This underscores the vast importance and dependence upon this antibody class in immunity against relapsing fever *Borrelia* spp. and the importance of antibodies, in general, in immunity to *Borrelia*.

Antibodies are also involved in the mechanism of antigenic variation utilized by these spirochetes (Stoenner et al. 1982). Initially, it was observed that fluorescent antiserum to *B. hermsii* was serotype-specific (Coffey and Eveland 1967a). This was expanded by the finding that the antibody response in relapsing fever develops specific to the dominant serotype of a spirochetemia peak, eliminating this population, and allowing for a different serotype, previously present in very low numbers, to thrive. In this manner the antibody/relapse cycle perpetuates in the infection (Stoenner et al. 1982). These serotype-specific antibody responses occur due to the fact that the majority of relapsing fever spirochetes at any one time are comprised of those that predominantly express one Vmp, the major antigens of this disease (Barbour et al. 1982; Barstad et al. 1985).

An antibody class response profile has been observed in *B. duttonii* infection of BALB/c mice (Yokota et al. 1997). A rapid antibody response was seen as IgM first emerged, followed by IgG2a and IgG3, after which the subsequent emergence of IgG1 and IgG2b occurred. It was assessed that IgG3 and IgM were the most important antibody classes in this response, as they were both shown to be passively protective against *B. duttonii* infection (Morshed et al. 1993; Yokota et al. 1997).

An interesting role for antibodies in relapsing fever infection was seen in *B. duttonii* infection (Morshed et al. 1993). Here it was shown that antibody transfer from mother to young ddY mice through milk or yolk sac route was effective for protection of offspring. IgG1 and IgG2a were shown to be predominantly transferred in milk, low levels of IgG2b were transferred via both routes, IgG3 was predominantly transferred through the yolk sac, and IgM was not transferred at all. Furthermore, IgG3 was the only class shown to be protective for offspring, suggesting that this antibody class may be important for protection in this aspect of the disease (Morshed et al. 1993).

The protection afforded by antibodies in *Borrelia* infection also includes phagocytosis of opsonized spirochetes (Benach et al. 1984; Filgueira et al. 1996; Lovrich et al. 2005; Montgomery et al. 1993, 1994, 2002; Montgomery and Malawista 1996; Rittig et al. 1992, 1998; Spagnuolo et al. 1982). It appears that *Borrelia* cannot survive the bacterial-killing weaponry of professional phagocytes. The general consensus is that spirochetes are killed by the cells that ingest them and that there are no known defense mechanisms against phagocytosis. The predominant mechanism of spirochete ingestion appears to be coiling phagocytosis (Rittig et al. 1992). This is a mechanism in which the phagocyte forms one pseudopod and coils around the pathogen until it is brought to the cell within a phagosome (Horwitz 1984). Destruction of *B. burgdorferi* by this mechanism occurs without the involvement of lysosomes and is influenced by NO and O_2 radical formation (Modolell et al. 1994;

Rittig et al. 1992). This type of phagocytosis is capable of spirochete discrimination as different removal frequencies were induced depending on the invading spirochete (Rittig et al. 1998). The Rho GTPase, CDC42Hs, is important for *B. burgdorferi* removal since its inhibition reduced pseudopod formation. Wiskott-Aldrich Syndrome protein, Arp2/3 (actin polymerization regulators), and f-actin are recruited to coiling pseudopods, suggesting the importance of these components in formation of pseudopods for *B. burgdorferi* removal (Linder et al. 2001). Opsonization of *B. burgdorferi* by antibodies plays an important role as the rate of coiling phagocytosis, NO and O_2 radical synthesis, and spirochete killing were all enhanced when borreliae were opsonized (Modolell et al. 1994; Rittig et al. 1992). In this aspect of the host response, antibodies are important as they bridge the innate and acquired immune systems by opsonizing borreliae for removal via Fc receptors on phagocytes. Considering that *Borrelia* can evade complement deposition, opsonization by antibodies followed by recognition of the Fc fragment may be the best method of phagocytic removal in *Borrelia* infection.

3.4 Antibody Responses and the Clinical Setting

Borrelia burgdorferi is a fastidious organism with a long cell division period. For this reason, bacterial cultures have never been used for diagnosis of active infections. Since a robust antibody response is the hallmark of the host response to *Borrelia*, serology (antibody reactivity to certain antigens) is the main method for diagnosis of Lyme disease. In studies prior to the discovery of *B. burgdorferi*, immunoglobulins and immunoglobulin complexes were shown to be useful markers supplementing the clinical diagnosis of Lyme disease (reviewed in Golightly 1997). Cryoglobulins and immune complexes were shown to be associated with disease (Hardin et al. 1979a, 1979b, 1984; Steere et al. 1977, 1979). These immune complexes, of the IgM and IgG classes, were found throughout disease and were specifically detected with the ^{125}I-C1q, C1q solid phase, or Raji cell assays (Hardin et al. 1979b). Antibodies were absolutely critical in the discovery of *B. burgdorferi*, as reactivity of patient sera to spirochetes and spirochetal lysate isolated from *Ixodes scapularis* ticks provided the first evidence linking this spirochete to the disease (Burgdorfer et al. 1982). Identification of common antigens between spirochetes isolated from patients and those isolated from ticks by MAbs solidified this identification (Barbour et al. 1983).

Since antibodies were instrumental in the identification of the Lyme disease agent, studies began investigating the use of antibodies in serodiagnosis of Lyme disease. Initially, an indirect immunofluorescence assay (IFA) utilizing whole spirochetes was used to screen antibodies in patient sera (Russell et al. 1984; Wilkinson 1984), but soon the enzyme-linked immunosorbent assay (ELISA) with whole *B. burgdorferi* lysate became the method of choice (Russell et al. 1984; Wilkinson 1984). However, the ELISA had specificity problems, especially with cross-reactivity of patient sera to other spirochetes and even other bacteria, and with sensitivity during the early stages of the disease (Karlsson et al. 1989; Sood et al. 1993). An immunoblot procedure containing whole *B. burgdorferi* lysate was added to the ELISA for a two-tiered ELISA-immunoblot serodiagnosis protocol (Davidson et al. 1996; Johnson et al. 1996; Ledue et al. 1996; Pachner and Ricalton 1992; Stanek 1991). Distinct banding pattern criteria were developed for the immunoblot, which allowed the two-tiered test to improve serologic sensitivity and specificity. The two-tiered test remains the preferred method for the laboratory diagnosis of Lyme disease. This diagnostic algorithm is recommended by the CDC (Johnson et al. 1996; Ledue et al. 1996; U.S. Department of Health and Human Services 1995) and is based on the banding criteria developed by testing well-defined patient populations (Dressler et al. 1993; Engstrom et al. 1995). Alternative serodiagnosis methods have been developed, such as the borreliacidal antibody test (Callister et al. 1991, 1993). This test measured bactericidal activity of patient sera against *B. burgdorferi*. If a certain level of bactericidal activity was observed, patients were considered seropositive for Lyme disease (Callister et al. 1991, 1993). This method was further refined using flow cytometry (Callister et al. 1994, 1996; Creson et al. 1996).

More recently, a very specific ELISA for Lyme disease, the C_6 Lyme test, was developed. This test utilizes C_6, a peptide derived from invariant region 6 (IR_6) of the Vmp-like sequence expressed (VlsE) antigen (Liang et al. 1999a, 1999b; Liang and Philipp 1999; Zhang et al. 1997). The ELISAs were initially performed with the entire VlsE protein, but were soon after cut down to the C_6 fragment, as its size allows for easy production and as IR_6 is the most conserved region in VlsE (Liang et al. 2000a). Currently, this seems to be the best serodiagnostic test for Lyme disease because it is very specific, sensitive, and rapid, allowing for the omission of a confirmatory test (Bacon et al. 2003; Embers et al. 2007; Jansson et al. 2005; Lawrenz et al. 1999; Liang et al. 1999b, 2000b; Magnarelli et al. 2002; Mogilyansky et al. 2004). Furthermore, there is evidence that this assay can also be used to stage the disease as a decline in antibody levels to C_6 correlates with successful treatment (Philipp et al. 2001; Philipp et al. 2005). This test underscores the tight relationship between the observations of basic science and the clinical applicability of such observations. The characterization of VlsE as the antigen associated with antigenic variation in *B. burgdorferi* had nothing to do with serologic diagnosis (Liang et al. 2000a; Zhang et al. 1997), yet the antigen has been adapted successfully to the clinical setting, reflecting the obvious benefits of translational research. The fact that serodiagnosis is the main method of diagnosis for *Borrelia* infection underscores the scope and importance of the antibody response in this infection and adds one more important role to the antibody repertoire of function.

Antibodies are important in diagnosis of relapsing fever, as well. Recently, an antigen specific to relapsing fever was discovered and used to differentiate Lyme disease from relapsing fever in serodiagnosis, as Vlps and Vsps are far too variable for this purpose. The glycerophosphodiester phosphodiesterase (GlpQ) of *B. hermsii* and *B. recurrentis* was identified as a borrelial antigen specific to relapsing fever and not present in *B. burgdorferi* or *Treponema* (Porcella et al. 2000; Schwan et al. 1996, 2003). Anti-GlpQ antibodies were also reactive against *B. crocidurae*, *B. parkeri*, *B. turicatae*, *B. coriaceae*, and *B. anserina* indicating that this antigen can be used to distinguish louse-borne and a number of tick-borne relapsing fever species from Lyme disease and syphilis in serodiagnosis (Porcella et al. 2000). Interestingly, a *glpQ* ortholog was identified in *B. lonestari*, the suspected agent of southern tick-associated rash illness (STARI), and so this may play a role in differentiating this pathogen from other borreliae in serodiagnosis (Bacon et al. 2004).

3.5 Antibody Responses Associated with Specific Tissue Manifestations of Borrelia Infection

Intrathecal antibodies, produced locally, have been shown to be induced in the neuroborreliosis of Lyme disease (reviewed in Garcia-Monco and Benach 1995; Garcia-Monco and Benach 1997; Halperin et al. 1989; Hansen et al. 1990; Kaiser and Rauer 1998; Li et al. 2006; Martin et al. 1988; Murray et al. 1986; Steere et al. 1990; Wilske et al. 1986). Accumulation of B cells in cerebrospinal fluid (CSF) correlated with the intrathecal antibody response, suggesting compartmentalization of the B cell response in the nervous tissue (Baig et al. 1989; Beuche et al. 1992). Immunoglobulin G, IgM, and IgA classes with reactivity against OspA, OspB, and OspC have all been associated with neuroborreliosis in CSF (Baig et al. 1989; Schutzer et al. 1997). In patients where neuroborreliosis occurred early in disease, an intrathecal IgM response to OspC was observed, consistent with the antigenic profile of *B. burgdorferi* (Schutzer et al. 1997). Intrathecal antibodies are markers of neurologic disease in patients, and have been used effectively for diagnosis and staging of the infection. While intrathecal antibodies are not unique to borrelia infections of the CNS, these are particularly prominent in neuroborreliosis.

Some evidence suggests that intrathecal antibody responses may not occur during infection with relapsing fever *Borrelia*. In the first 2 weeks of *B. crocidurae* infection in C57BL/6 mice, there was minimal B and T cell infiltration in the brain (Andersson et al. 2007). Infiltrates predominantly consisted of innate immune cells (Andersson et al. 2007). This, however, was one case and may not be representative of other relapsing fever infections in which B cells and plasma cells comprised a prominent portion of infiltrates (Garcia-Monco et al. 1997). Interestingly, differences were observed in the pathogenesis of two different *B. turicatae* serotypes during infection of CB-17 and C3H SCID mice (Cadavid et al. 1994; Pennington et al. 1997). One serotype (serotype A) exhibited central nervous system (CNS) invasion early in infection, and an increased ability to penetrate HUVEC monolayers. The other serotype (serotype B), however, exhibited a preference for joint infiltration as it caused enlarged and reddened joints, severe arthritis as determined by histology, and impaired the ability of mice to balance on a walking bar (Cadavid et al. 1994). Furthermore, serotype B achieved higher levels in the blood and joints (Pennington et al. 1997), whereas serotype A achieved higher levels in the leptomeninges and brain (Cadavid et al. 2001; Sethi et al. 2006). This is of interest as antibodies are linked to the production of different relapsing fever serotypes (discussed in Sect. 4). In this scenario, the host antibody response may eliminate one relapsing fever serotype with a preference for a certain tissue, allowing for the production of an alternate serotype and thus preference for a different tissue. This becomes apparent as residual brain infection with *B. turicatae* serotype A occurred in about 20% of immunocompetent C57BL/6 mice (Cadavid et al. 2006). This suggests that the other 80% of mice were infected with *B. turicatae* that were forced to seroconvert due to the host antibody response, and thus lost their neurotropism. These findings indirectly demonstrate a role for host serum antibodies in relapsing fever *Borrelia* tissue preference.

In initial observations of chronic Lyme arthritis, serum IgM was seen in disease and IgG associated with remission (Steere et al. 1979). Elevated IgA immune complexes, in the form of rheumatoid factors, were seen in approximately 25% of patients with Lyme arthritis in one study, but did not seem to be associated with disease (Axford et al. 1999). Recently it was shown that IgG is in abundance in chronic Lyme arthritis synovial fluid, and appears to be present due to antigen-driven selection (Ghosh et al. 2005). The importance of serum IgG in Lyme arthritis was further demonstrated as the presence of the neonatal Fc receptor (FcRn), an important receptor in recycling IgG and extending its serum half-life, was shown to contribute to the resolution of arthritis (Crowley et al. 2006). This was achieved by demonstrating that FcRn$^{-/-}$ mice infected with *B. burgdorferi* had decreased anti-*B. burgdorferi* bactericidal IgG and increased ankle swelling (Crowley et al. 2006). Transfer of immune serum to *B. burgdorferi* infected SCID mice has also been shown to mediate arthritis remission (Barthold et al. 1996, 1997, 2006).

Removal of opsonized *B. burgdorferi* by phagocytes, while beneficial in the host response, has been associated with Lyme arthritis pathology, as there is evidence for the involvement of macrophages in the induction of severe destructive Lyme arthritis (Du Chateau et al. 1996). Interaction of macrophages with T cells was shown to induce severe destructive Lyme arthritis and these cells synergistically induced the condition (DuChateau et al. 1997, 1999). So, as there are benefits in the phagocytic removal of *Borrelia*, it seems as though enhancement of *B. burgdorferi* phagocytosis also plays a role in the promotion Lyme arthritis. These findings indicate the involvement and importance of antibodies not only in immunity to *Borrelia* in the blood but also in the resolution and possibly the pathology of tissue manifestations.

3.6 Antibody-Mediated Autoimmunity in Borrelia Infection

Autoimmunity has long been sought as a major reason for the pathogenesis of chronic infections. In particular, the chronic course of the spirochetoses, whether syphilis or the borrelioses, has prompted investigations into the possible role of autoimmunity to explain the long-term infectious process. Antibodies elicited by *Borrelia* infection have raised the possibility of autoimmunity in the neurological and, more recently, the arthritic manifestations of Lyme disease (reviewed in Garcia-Monco et al. 1995; reviewed in Sigal 1997). Manifestations of acute and chronic neuroborreliosis include cranial neuropathy resulting in facial palsy (Bell's palsy) and axonopathy in peripheral neurologic Lyme disease and central nervous system involvement, including encephalopathy (reviewed in Garcia-Monco et al. 1995;

reviewed in Halperin 2003, 2005; Halperin and Golightly 1992; Halperin et al. 1987, 1989, 1991; reviewed in Sigal 1997; reviewed in Steere 2001). *Borrelia burgdorferi* is rarely seen in nervous system tissue biopsies and has never been seen in peripheral nerve tissue biopsies in the mouse model. This suggests that something other than infection, possibly autoimmunity, may contribute to this manifestation (reviewed in Duray 1989; reviewed in Sigal 1997). An early indication of autoimmune involvement in neuroborreliosis was the finding that serum from neurologic Lyme patients contained an abundance of anti-*B. burgdorferi* IgM, autoreactive to human peripheral nerve axons (Sigal and Tatum 1988). Furthermore, a murine monoclonal IgG (H9724) specific for *B. burgdorferi* flagellin$_{213-224}$ was shown to be reactive against human nerve axons and immature neuritic processes, and, more specifically, heat shock protein (HSP)-60 of neuroblastoma cells (Fikrig et al. 1993; Sigal 1993; Sigal and Tatum 1988; Yu et al. 1997). Patient sera was also shown to be reactive against HSP-60, indicating this host protein as a primary autoimmune target of antibodies in neurologic Lyme disease (Sigal 1993). The effects of H9724 were shown to prevent axonal formation, even in the presence of neuronal growth factor (NGF) and basic fibroblast growth factor (bFGF) (Sigal and Williams 1997). This suggests that autoreactive antibodies may be causing the prevention of axon formation associated with neurological Lyme disease. Autoreactive antibodies have also been observed in the central nervous manifestations of Lyme disease (Schluesener et al. 1989), as IgM antibodies cross-reactive against gangliosides and nonprotein antigenic fractions have been induced by *B. burgdorferi* infection (reviewed in Garcia-Monco et al. 1995; Garcia Monco et al. 1993; Wheeler et al. 1993). Antibodies to nonprotein components of *B. burgdorferi* were detected by thin-layer chromatography of lipid or glycolipid fractions of these organisms (Wheeler et al. 1993). The *Borrelia* glycolipids have been identified as cholesteryl 6-O-acyl-β-d-galactopyranoside (*B. burgdorferi* glycolipid I [Bb GL-I]) and 1,2-di-O-acyl-3-O-α-d-galactopyranosyl-sn-glycerol (*B. burgdorferi* glycolipid II [Bb GL-II]) (Ben-Menachem et al. 2003; Schroder et al. 2003). Intrathecal antibodies of the IgG, IgM, and IgA classes from patient cerebrospinal fluid were also shown to be autoreactive, as they bound myelin basic protein and different neurofilament proteins (Kaiser 1995). In addition, anti-OspA antibodies have been shown to be cross-reactive with neurons in the brain, spinal cord, and dorsal root ganglia, implicating autoantibodies induced by *B. burgdorferi* as potential contributors to the pathology of the disease (Alaedini and Latov 2005).

Treatment-resistant Lyme arthritis (TRLA) is a condition observed in approximately 10% of Lyme arthritis patients, characterized by continuous joint inflammation that does not resolve upon therapy, and, moreover, is not associated with active *B. burgdorferi* infection (reviewed in Guerau-de-Arellano and Huber 2002; reviewed in Steere et al. 2001; reviewed in Weinstein and Britchkov 2002). T_{H1}-cell cross-recognition of OspA$_{165-173}$ and human lymphoid/myeloid adhesion molecule LFA-1 has previously been associated with TRLA, making a strong case for autoimmunity in the pathology of this condition (Gross et al. 1998; Guerau-de-Arellano and Huber 2002; Steere et al. 2003; Trollmo et al. 2001). Subsequent to this discovery, the involvement of autoreactive antibodies was suggested to contribute to this condition (Guerau-de-Arellano and Huber 2002). Antisera reactivity to OspA and OspB has consistently been observed in patients with TRLA (Chen et al. 1999; Kalish et al. 1993). Indeed, through experiments that made use of single-chain variable fragments (scFvs) derived from IgG found in patient synovium, antibody cross-reactivity between OspA and host cytokeratin (CK)-10 was observed (Ghosh et al. 2005, 2006). Thus, there seems to be a role for autoantibodies in the pathology of TRLA that has not yet been completely elucidated. The role for autoimmunity in the pathogenesis of *Borrelia* would certainly explain the many enigmatic observations that have been made in the natural human infection as well as in the experimental murine setting.

4 Unique Properties of Antibodies in *Borrelia* Infection

4.1 Immune Pressure and Development of Escape Mutants

In addition to their traditional roles in host defense, antibodies are the effectors in antigenic variation of *Borrelia*. The specificity of antibodies is such that only certain epitopes are recognized (serotype specificity) and organisms bearing those epitopes are eliminated, allowing other serotypes to have unrestricted growth. Thus, antibodies appear to impose such stress on *Borrelia* that they cause the appearance of surface antigen escape mutants (Coleman et al. 1992, 1994; Hodzic et al. 2005; Liang et al. 2002; Sadziene et al. 1994). In C3H/HeN mice, it was seen at 17 d.p.i. that anti-OspC antibodies emerge coincident with the appearance of OspC variants (Liang et al. 2002). Moreover, the variants re-expressed OspC in culture and naïve mice. In SCID mice, OspC was persistently expressed by *B. burgdorferi*; however, passively administered anti-OspC MAb eliminated these spirochetes, which then re-emerged when MAb was taken away (Liang et al. 2002). OspC mRNA was also greatly decreased under immune pressure in vivo, and BBF01 and vlsE were concomitantly upregulated (Liang et al. 2004). In SCID mice, OspA transcription was regulated by the presence of mouse serum, IgG, or IgM, suggesting that Osp expression is greatly influenced by host antibodies (Hodzic et al. 2005). Furthermore, constitutive expression of OspC by the *B. burgdorferi* flagellar gene (FlaB) promoter resulted in clearance of spirochetes in vivo (SCID mice), when anti-OspC MAb was administered (Xu et al. 2006). This indicates that differential Osp expression in response to the great pressure imposed by host antibodies is an important persistence mechanism of *B. burgdorferi* during infection (Xu et al. 2006).

4.2 Antibody Interactions Within the Tick Vector

Antibodies against *Borrelia* also have an effect on spirochetes in the tick vector (see Fig. 4). This was clearly demonstrated in the work that led to the creation of a Lyme disease OspA subunit vaccine. It was observed early on that OspA elicited

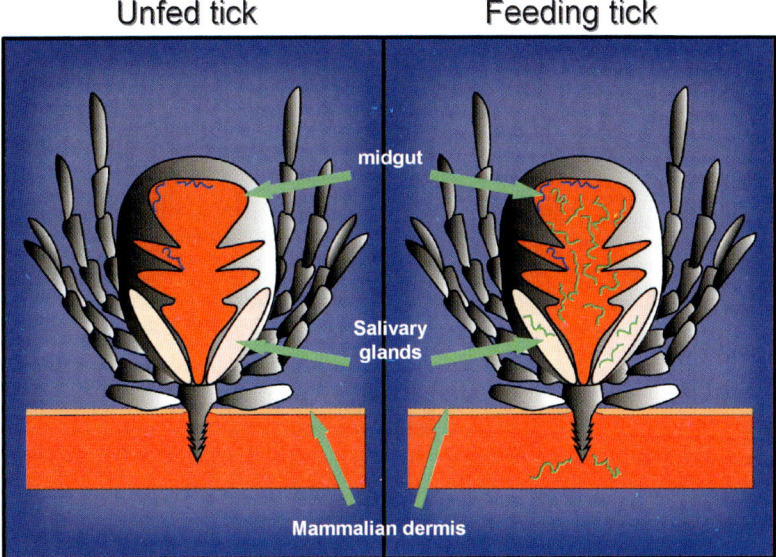

Fig. 4 Tick transmission of *Borrelia*. Depicted in the diagram is the tick-transmission of Lyme disease *Borrelia* via a hard-bodied *Ixodes* tick. In an unfed tick, spirochetes are present in very low numbers in the tick midgut (23°C, pH ~7.5) and predominantly express the surface lipoprotein OspA (depicted in blue). Upon feeding, mammalian blood enters the midgut and dramatically changes the environment (35°C, pH <7.0). The new environment allows for the spirochetes to replicate to greater numbers. As the tick feeds, spirochetes migrate to the salivary glands where they now predominantly express OspC (depicted in green) in place of OspA. While in the salivary glands, spirochetes can be transmitted to the mammal via tick saliva. The tick must feed for approximately 48 hours for efficient transmission of Lyme disease *Borrelia*. Tick-transmission of relapsing fever *Borrelia* occurs similarly via the soft-bodied *Ornithodoros* ticks. However, these ticks feed for very short periods of time (less than a half hour) and can efficiently transmit the infecting spirochetes in this time frame. Relapsing fever *Borrelia* express different lipoproteins than do Lyme disease *Borrelia*, as well

protective antibodies, but exactly how the antibodies prevented infection became apparent when the effect of the antibody was shown to occur in the tick vector stage (Fikrig et al. 1990, 1992b). Outer surface protein A is highly expressed by spirochetes in the tick gut (as it is in culture) and downregulated upon blood feeding by the tick. Increased expression of OspC was noted more or less simultaneously with incoming host blood (see Fig. 4) (Schwan et al. 1995). Antibodies to OspA blocked transmission of spirochetes to mammalian hosts by acting within a narrow window of time, when blood meal, where host antibodies would be present, is taken up by the ticks (de Silva et al. 1996). As possible variation in surface Osps became a question, it was shown that OspA antiserum was more efficient at spirochete killing in the presence of complement in vitro, allowing for destruction of escape variants (Nowling and Philipp 1999; Sole et al. 1998). However, complement is likely inactivated in the tick and was shown to be dispensable for efficient spirochete blockage, as anti-OspA antibodies were not required to kill *B. burgdorferi* in the tick gut to exert their transmission-blocking effects (de Silva et al. 1999; Gipson and de Silva 2005; Rathinavelu et al. 2003). The fact that antibodies are the sole purveyors in blocking *Borrelia* transmission from the tick is direct evidence of their power and versatility. The protection afforded by antibodies in response to the OspA vaccine was quite novel at the time of its discovery, and was remarkably effective. In clinical trials it was 72% and 92% effective with and without adjuvant, respectively (Sigal et al. 1998; Steere et al. 1998). A number of concerns regarding the vaccine resulted in its removal from the market (Akin et al. 1999; Croke et al. 2000; Gross et al. 1998; Rose et al. 2001). Studies are currently underway for the design of a second-generation OspA vaccine that eliminates the N-terminal portion of the protein implicated in autoimmunity (Koide et al. 2005; Nassal et al. 2005; Scheckelhoff et al. 2006; Willett et al. 2004). It has been noted recently that B cell inhibitory proteins exist within *Ixodes ricinus* and *Hyalomma asiaticum asiaticum* tick vectors, but what this means to spirochete transmission is not quite clear (Hannier et al. 2004; Yu et al. 2006). The method of protection of anti-OspA antibodies underscores the importance of interactions that may occur at the tick–host interface, and widens the potential roles that antibodies may play in immunity.

4.3 *The Presence and Importance of Complement-Independent, Bactericidal Antibodies*

The array of functions performed by antibodies was expanded with the discovery of *Borrelia*-specific antibodies that had complete bactericidal capability in the absence of complement. Initial indications of this unique capability were seen in vivo as complement was shown to be dispensable for efficient immunity against relapsing fever and Lyme disease *Borrelia*. Mice deficient in C5, required for the lytic effects of complement, had no defect in removal of relapsing fever agent *B. turicatae* as compared to controls (Newman and Johnson 1981). The irrelevance of complement in the host response to relapsing fever was further demonstrated as C5-, C3-, and C1q-deficient (defective for complement opsonization) mice efficiently cleared the spirochetemia caused by an uncultivable relapsing fever *Borrelia* originally isolated from patients (Anda et al. 1996; Connolly and Benach 2001; Connolly et al. 2004). In *B. burgdorferi* murine infection, depletion of complement with cobra venom factor suggested that complement may not be necessary for an efficient host response against the spirochete; however, at times during the infection complement was needed (Schmitz et al. 1991). Further in vivo studies with C5-deficient mice conclusively showed that lytic complement was unnecessary for efficient immunity against *B. burgdorferi* (Bockenstedt et al. 1993).

The first anti-*Borrelia* complement-independent MAb, CB2 (IgG1), was specific to OspB in *B. burgdorferi* and shown to exert bactericidal effects in vitro (Coleman et al. 1992). The CB2:OspB interaction was shown to be dependent on lysine 253 in OspB, without which binding does not occur and bactericidal activity is abrogated (Coleman et al. 1994). Importantly, the Fab fragment of CB2 was shown to be bactericidal in the absence of complement in vitro, proving that agglutination did not contribute to activity, as these molecules are monomeric. The damage imparted to *B. burgdorferi* after CB2 binding included destabilization and destruction of the outer membrane, likely resulting in lysis of the spirochete (Coleman et al. 1992; Escudero et al. 1997). Another complement-independent, bactericidal MAb specific for OspB, H6831(IgG2a), was discovered and shown to be functionally identical to CB2, even dependent upon the same residue, lysine 253, for binding and activity (Sadziene et al. 1994). Other *B. burgdorferi*-specific, complement-independent, bactericidal antibodies with reactivity to OspA, OspB, and P39 have also been discovered, but have not been investigated insofar as to completely characterize them and their bactericidal ability (Aydintug et al. 1994; Barbour and Bundoc 2001; Ma et al. 1995; Scriba et al. 1993).

Complement-independent, bactericidal MAbs against relapsing fever *Borrelia* spp. have also been discovered and investigated, and seem to function similarly to those *B. burgdorferi*-specific MAbs previously discussed (Barbour and Bundoc 2001; Connolly et al. 2004; Sadziene et al. 1994). Two such MAbs are CB515 and H4825, of IgM and IgG2a classes, respectively (Connolly et al. 2004; Sadziene et al. 1994). One of these MAbs, CB515, has shown in vivo significance, as it was derived from a polyclonal complement-independent IgM response to relapsing fever and was shown to passively protect B cell-deficient, C5-deficient, and wild type C57BL/6 mice (Connolly and Benach 2001; Connolly et al. 2004). This IgM was shown to be reactive against a Vsp and caused similar damage to that caused by CB2, H6831, and H4825. This included the disruption of the spirochetal outer membrane, allowing for exposure of periplasmic flagella and the induction of severe membrane blebbing, likely resulting in lysis of the pathogen (Connolly et al. 2004). It is tempting to speculate that the in vivo complement-independent IgM response against relapsing fever from which CB515 is derived, and other complement-independent IgM MAbs against relapsing fever, may be derived from a T cell-independent, B1b/MZ B cell response, shown to be critical for resolution of relapsing fever spirochetemia. As B1b and MZ B cells are both IgM secreting cells and as both these B cell subsets and complement-independent IgM antibodies were shown to be critical for spirochetemia clearance in relapsing fever, they seem to correlate. However, production of complement-independent IgM antibodies from these particular cells has not been directly observed experimentally (Alugupalli et al. 2003, 2004; Belperron et al. 2005). Therefore, whether or not these B cells are the source of the complement-independent IgM antibodies has not yet been conclusively proven.

The fact that these antibodies are capable of bactericidal activity is quite unique, as MAC formation via classical complement recruitment has traditionally been thought of as the mechanism whereby antibodies can exert a bactericidal effect without cellular assistance. However, complement-dependent antibodies against *Borrelia* exist (discussed in Sect. 2.1), and generally it is thought that complement-dependent and complement-independent antibodies work in concert to make possible efficient immunity against *Borrelia* infection (Aydintug et al. 1994; Bockenstedt et al. 1993; Ma et al. 1995; Schmitz et al. 1991). Where have the complement-independent antibodies come from, and are they present in immunity against any infection? The answers to these questions are not known; however, there may be some clues as to why these antibodies exist in the *Borrelia* immune response and are particularly prominent and effective in the response to relapsing fever. *Borrelia* have the ability to evade complement-mediated destruction via the alternative pathway (discussed in Sect. 2.2). Since they are complement-resistant, this may increase the importance of complement-independent antibodies. However, this does not rule out possibility of *B. burgdorferi* destruction induced by complement via the classical pathway, on which complement-dependent antibodies rely (Kochi and Johnson 1988; Kochi et al. 1991, 1993). The recent finding that relapsing fever *Borrelia* are also able to bind C4BP, suggests that these spirochetes are capable of evasion of the classical complement pathway (Meri et al. 2006). Interestingly, this may render complement completely ineffective against relapsing fever *Borrelia*, allowing complement-independent, bactericidal antibodies to come into prominence in immunity against this infection. This may explain why complement-independent, bactericidal antibodies are so critical for relapsing fever clearance. These connections, however, have not been made experimentally and are, thus far, speculation.

Although the bactericidal mechanism has not been elucidated, there have been studies that investigated the function of some of these complement-independent, bactericidal antibodies. Polar blebbing was shown to be induced in *B. burgdorferi* upon CB2 Fab:OspB immune complex formation (Escudero et al. 1997). This was followed by OspA colocalization, and spheroplast induction causing outer membrane destabilization and, ultimately, lysis of the spirochetes (Escudero et al. 1997). Bactericidal function of the CB2 Fab was also shown to be dependent on the presence of Mg^{2+} and Ca^{2+}, but how these ions may be specifically involved in the process remains unclear (Escudero et al. 1997). Upon binding, CB2 also causes a change in OspB that is reflected in a difference in the susceptibility of the antigen to certain proteases (Katona et al. 2000). Specific structural changes, in the form of a disordered secondary structure, were observed in OspB upon H6831 Fab binding, after investigation of the immune complex crystal structure (Becker et al. 2005). The fact that OspB undergoes changes after complement-independent, bactericidal MAb binding suggests that these antibodies may induce a process in *Borrelia* upon binding. In fact, upregulation of *blyA* and *blyB* phage holins system genes, of circular plasmid (cp) 32 in *B. burgdorferi*, was observed upon CB2 Fab binding to OspB (Anderton et al. 2004). It is not known if complete phages were assembled upon binding. However, the BlyA protein is known to be membrane-interactive and, upon *blyA* and *blyB* upregulation, phage induction has been shown to occur (Damman et al. 2000). Interestingly, phage release was associated with spontaneous lysis of *B. hermsii* (Barbour and Hayes 1986), which contain homologous holins genes to *B. burgdorferi* (Stevenson et al. 2000). Moreover, phage production in *Borrelia* has been shown to be induced in response to environmental stress (Hayes et al. 1983; Neubert et al. 1993). This lends support to the idea that complement-independent MAb binding to *Borrelia* may result in phage formation, possibly contributing to the bactericidal effect of this interaction.

Other possibilities include the utilization of the antibody-catalyzed water-oxidation pathway (ACWOP) (Wentworth et al. 2000). This is a process proposed to be universal among all antibodies regardless of specificity. In this process, antibodies can utilize singlet oxygen ($^1O_2^*$) and water in order to produce reactive oxygen species such as hydrogen peroxide (H_2O_2), dihydrogen trioxide (H_2O_3), and ozone (O_3) (Wentworth et al. 2000, 2003). Production of such species would allow the antibodies to become intrinsically bactericidal (Wentworth et al. 2000, 2002, 2003). It is proposed that there is an active site located in all antibodies within the interfacial region of the constant and variable regions, which catalyzes this reaction (Zhu et al. 2004). This is proposed as residues within both the constant and variable regions of Fab fragments were modified after exposure to UV radiation or H_2O_2. While some of these modifications were inconsistent, a residue within the constant region, Trp^{L163}, was consistently and extensively modified when exposed to UV or H_2O_2. Modification of Trp^{L163} occurred in two different Fabs, and this residue is highly conserved among murine antibodies, suggesting that it is important in the utilization of this reaction (Zhu et al. 2004). In a host, it is thought that this reaction may be allowed to begin as $^1O_2^*$ is released from phagocytes due to oxidative burst (Babior et al. 2003). It is unlikely that these *Borrelia* complement-independent MAbs utilize this as their method of killing, as a source of $^1O_2^*$, such as a phagocyte, was not included in their in vitro bactericidal assays. Irrelevant MAb controls used in these experiments did not affect *Borrelia* either, further suggesting that conditions for this pathway probably did not exist in these experiments (Coleman et al. 1992; Connolly et al. 2004; Ma et al. 1995; Sadziene et al. 1994). It has recently been shown that antibodies can catalyze the antibody-catalyzed water-oxidation pathway through utilization of riboflavin (vitamin B2) (Nieva et al. 2006). This may possibly be a mechanism utilized by the anti-*Borrelia* complement-independent MAbs, as riboflavin is a component of the Barbour-Stoenner-Kelly (BSK)-H medium in which the in vitro bactericidal assays were conducted. However, as stated previously (Sect. 2.1), there are anti-*Borrelia* antibodies that are nonbactericidal in the absence of complement and the presence of a source of riboflavin (serum or BSK-H) (Aydintug et al. 1994; Lovrich et al. 1991; Ma et al. 1995; Munson et al. 2000, 2006; Nowling and Philipp 1999; Remington et al. 2001; Rousselle et al. 1998). Thus utilization of riboflavin by antibodies from these sources seems unlikely, and casts doubt as to whether these particular antibodies utilize this pathway. Since Trp^{L163} of the constant region seems to be essential for utilization of this pathway, investigation of an scFv derived from an anti-*Borrelia* complement-independent, bactericidal MAb should indisputably indicate whether or not this pathway accounts for the unique bactericidal capability of these MAbs. An scFv would be an ideal vehicle for this investigation as these molecules contain no constant region and are wholly composed of the antibody variable region (reviewed in Holliger and Hudson 2005; Huston et al. 1988).

It is entirely possible that the anti-*Borrelia* complement-independent MAbs utilize a completely different and previously overlooked method to become bactericidal. As Fabs of some of these MAbs have been shown to exert bactericidal activity (Coleman et al. 1992; Sadziene et al. 1994), perhaps the answer may lie in their variable regions. Are the variable regions of these MAbs bactericidal on their own? It may be possible that part of the variable regions encode something not present in complement-dependent antibodies that allow these MAbs to have an innate bactericidal quality. An scFv would be useful in testing this theory, as well, since constant regions would not be present. If these MAbs are innately bactericidal, maybe they work in concert with changes that occur in their antigens and the borreliae upon binding, as discussed before. The question has, as yet, remained unanswered.

Acknowledgements We appreciate the assistance of Susan Malkiel, James Coleman, and Sean Connolly for helpful, critical reviews of the manuscript. We would also like to thank Rafal Tokarz for contributing Fig. 4 to the review.

References

Akin E, McHugh GL, Flavell RA, Fikrig E, Steere AC (1999) The immunoglobulin (IgG) antibody response to OspA and OspB correlates with severe and prolonged Lyme arthritis and the IgG response to P35 correlates with mild and brief arthritis. Infect Immun 67:173–181

Alaedini A, Latov N (2005) Antibodies against OspA epitopes of *Borrelia burgdorferi* cross-react with neural tissue. J Neuroimmunol 159:192–195

Alitalo A, Meri T, Lankinen H, Seppala I, Lahdenne P, Hefty PS, Akins D, Meri S (2002) Complement inhibitor factor H binding to Lyme disease spirochetes is mediated by inducible expression of multiple plasmid-encoded outer surface protein E paralogs. J Immunol 169:3847–3853

Alitalo A, Meri T, Chen T, Lankinen H, Cheng ZZ, Jokiranta TS, Seppala IJ, Lahdenne P, Hefty PS, Akins DR, Meri S (2004) Lysine-dependent multipoint binding of the *Borrelia burgdorferi* virulence factor outer surface protein E to the C terminus of factor H. J Immunol 172:6195–6201

Alugupalli KR, Gerstein RM, Chen J, Szomolanyi-Tsuda E, Woodland RT, Leong JM (2003) The resolution of relapsing fever borreliosis requires IgM and is concurrent with expansion of B1b lymphocytes. J Immunol 170:3819–3827

Alugupalli KR, Leong JM, Woodland RT, Muramatsu M, Honjo T, Gerstein RM (2004) B1b lymphocytes confer T cell-independent long-lasting immunity. Immunity 21:379–390

Anda P, Sanchez-Yebra W, del Mar Vitutia M, Perez Pastrana E, Rodriguez I, Miller NS, Backenson PB, Benach JL (1996) A new *Borrelia* species isolated from patients with relapsing fever in Spain. Lancet 348:162–165

Anderson JF, Magnarelli LA, McAninch JB (1988) New *Borrelia burgdorferi* antigenic variant isolated from *Ixodes dammini* from upstate New York. J Clin Microbiol 26:2209–2212

Anderson JF, Magnarelli LA, LeFebvre RB, Andreadis TG, McAninch JB, Perng GC, Johnson RC (1989) Antigenically variable Borrelia burgdorferi isolated from cottontail rabbits and *Ixodes dentatus* in rural and urban areas. J Clin Microbiol 27:13–20

Andersson M, Nordstrand A, Shamaei-Tousi A, Jansson A, Bergstrom S, Guo BP (2007) In situ immune response in brain and kidney during early relapsing fever borreliosis. J Neuroimmunol 183:26–32

Anderton JM, Tokarz R, Thill CD, Kuhlow CJ, Brooks CS, Akins DR, Katona LI, Benach JL (2004) Whole-genome DNA array analysis of the response of *Borrelia burgdorferi* to a bactericidal monoclonal antibody. Infect Immun 72:2035–2044

Anguita J, Rincon M, Samanta S, Barthold SW, Flavell RA, Fikrig E (1998) Borrelia burgdorferi-infected, interleukin-6-deficient mice have decreased Th2 responses and increased lyme arthritis. J Infect Dis 178:1512–1515

Axford JS, Rees DH, Mageed RA, Wordsworth P, Alavi A, Steere AC (1999) Increased IgA rheumatoid factor and V(H)1 associated cross reactive idiotype expression in patients with Lyme arthritis and neuroborreliosis. Ann Rheum Dis 58:757–761

Aydintug MK, Gu Y, Philipp MT (1994) *Borrelia burgdorferi* antigens that are targeted by antibody-dependent, complement-mediated killing in the rhesus monkey. Infect Immun 62:4929–4937

Babior BM, Takeuchi C, Ruedi J, Gutierrez A, Wentworth P Jr (2003) Investigating antibody-catalyzed ozone generation by human neutrophils. Proc Natl Acad Sci U S A 100:3031–3034

Bacon RM, Biggerstaff BJ, Schriefer ME, Gilmore RD Jr, Philipp MT, Steere AC, Wormser GP, Marques AR, Johnson BJ (2003) Serodiagnosis of Lyme disease by kinetic enzyme-linked immunosorbent assay using recombinant VlsE1 or peptide antigens of *Borrelia burgdorferi* compared with 2-tiered testing using whole-cell lysates. J Infect Dis 187:1187–1199

Bacon RM, Pilgard MA, Johnson BJ, Raffel SJ, Schwan TG (2004) Glycerophosphodiester phosphodiesterase gene (glpQ) of *Borrelia lonestari* identified as a target for differentiating *Borrelia* species associated with hard ticks (Acari:Ixodidae). J Clin Microbiol 42:2326–2328

Baig S, Olsson T, Link H (1989) Predominance of *Borrelia burgdorferi* specific B cells in cerebrospinal fluid in neuroborreliosis. Lancet 2:71–74

Baranton G, Postic D, Saint Girons I, Boerlin P, Piffaretti JC, Assous M, Grimont PA (1992) Delineation of *Borrelia burgdorferi* sensu stricto, *Borrelia garinii* sp. nov., and group VS461 associated with Lyme borreliosis. Int J Syst Bacteriol 42:378–383

Barbour AG (1990) Antigenic variation of a relapsing fever *Borrelia* species. Annu Rev Microbiol 44:155–171

Barbour AG (1991) Molecular biology of antigenic variation in Lyme borreliosis and relapsing fever: a comparative analysis. Scand J Infect Dis Suppl 77:88–93

Barbour AG, Bundoc V (2001) In vitro and in vivo neutralization of the relapsing fever agent *Borrelia hermsii* with serotype-specific immunoglobulin M antibodies. Infect Immun 69:1009–1015

Barbour AG, Hayes SF (1986) Biology of *Borrelia* species. Microbiol Rev 50:381–400

Barbour AG, Tessier SL, Stoenner HG (1982) Variable major proteins of *Borrellia hermsii*. J Exp Med 156:1312–1324

Barbour AG, Tessier SL, Todd WJ (1983) Lyme disease spirochetes and ixodid tick spirochetes share a common surface antigenic determinant defined by a monoclonal antibody. Infect Immun 41:795–804

Barbour AG, Dai Q, Restrepo BI, Stoenner HG, Frank SA (2006) Pathogen escape from host immunity by a genome program for antigenic variation. Proc Natl Acad Sci U S A 103:18290–18295

Barstad PA, Coligan JE, Raum MG, Barbour AG (1985) Variable major proteins of *Borrelia hermsii*. Epitope mapping and partial sequence analysis of CNBr peptides. J Exp Med 161:1302–1314

Barthold SW, Bockenstedt LK (1993) Passive immunizing activity of sera from mice infected with *Borrelia burgdorferi*. Infect Immun 61:4696–4702

Barthold SW, deSouza M, Feng S (1996) Serum-mediated resolution of Lyme arthritis in mice. Lab Invest 74:57–67

Barthold SW, Feng S, Bockenstedt LK, Fikrig E, Feen K (1997) Protective and arthritis-resolving activity in sera of mice infected with Borrelia burgdorferi. Clin Infect Dis 25 [Suppl 1]:S9–S17

Barthold SW, Hodzic E, Tunev S, Feng S (2006) Antibody-mediated disease remission in the mouse model of lyme borreliosis. Infect Immun 74:4817–4825

Baumgarth N, Herman OC, Jager GC, Brown LE, Herzenberg LA, Chen J (2000) B-1 and B-2 cell-derived immunoglobulin M antibodies are nonredundant components of the protective response to influenza virus infection. J Exp Med 192:271–280

Beck G, Habicht GS, Benach JL, Coleman JL (1985) Chemical and biologic characterization of a lipopolysaccharide extracted from the Lyme disease spirochete (*Borrelia burgdorferi*). J Infect Dis 152:108–117

Becker M, Bunikis J, Lade BD, Dunn JJ, Barbour AG, Lawson CL (2005) Structural investigation of *Borrelia burgdorferi* OspB, a bactericidal Fab target. J Biol Chem 280:17363–17370

Belperron AA, Dailey CM, Bockenstedt LK (2005) Infection-induced marginal zone B cell production of *Borrelia hermsii*-specific antibody is impaired in the absence of CD1d. J Immunol 174:5681–5686

Ben-Menachem G, Kubler-Kielb J, Coxon B, Yergey A, Schneerson R (2003) A newly discovered cholesteryl galactoside from *Borrelia burgdorferi*. Proc Natl Acad Sci U S A 100:7913–7918

Benach JL, Bosler EM, Hanrahan JP, Coleman JL, Habicht GS, Bast TF, Cameron DJ, Ziegler JL, Barbour AG, Burgdorfer W, Edelman R, Kaslow RA (1983) Spirochetes isolated from the blood of two patients with Lyme disease. N Engl J Med 308:740–742

Benach JL, Fleit HB, Habicht GS, Coleman JL, Bosler EM, Lane BP (1984) Interactions of phagocytes with the Lyme disease spirochete: role of the Fc receptor. J Infect Dis 150:497–507

Benach JL, Coleman JL, Golightly MG (1988) A murine IgM monoclonal antibody binds an antigenic determinant in outer surface protein A, an immunodominant basic protein of the Lyme disease spirochete. J Immunol 140:265–272

Bendelac A, Bonneville M, Kearney JF (2001) Autoreactivity by design: innate B and T lymphocytes. Nat Rev Immunol 1:177–186

Bergstrom S, Bundoc VG, Barbour AG (1989) Molecular analysis of linear plasmid-encoded major surface proteins, OspA and OspB, of the Lyme disease spirochaete *Borrelia burgdorferi*. Mol Microbiol 3:479–486

Beuche W, Siever A, Felgenhauer K (1992) Specific antigen binding by activated cerebrospinal fluid B lymphocytes in acute neuroborreliosis. J Neurol 239:322–326

Bockenstedt LK, Barthold S, Deponte K, Marcantonio N, Kantor FS (1993) *Borrelia burgdorferi* infection and immunity in mice deficient in the fifth component of complement. Infect Immun 61:2104–2107

Bockenstedt LK, Fikrig E, Barthold SW, Flavell RA, Kantor FS (1996) Identification of a *Borrelia burgdorferi* OspA T cell epitope that promotes anti-OspA IgG in mice. J Immunol 157:5496–5502

Bockenstedt LK, Hodzic E, Feng S, Bourrel KW, de Silva A, Montgomery RR, Fikrig E, Radolf JD, Barthold SW (1997) *Borrelia burgdorferi* strain-specific Osp C-mediated immunity in mice. Infect Immun 65:4661–4667

Bolz DD, Sundsbak RS, Ma Y, Akira S, Weis JH, Schwan TG, Weis JJ (2006) Dual role of MyD88 in rapid clearance of relapsing fever *Borrelia* spp. Infect Immun 74:6750–6760

Brandt ME, Riley BS, Radolf JD, Norgard MV (1990) Immunogenic integral membrane proteins of *Borrelia burgdorferi* are lipoproteins. Infect Immun 58:983–991

Brooks CS, Vuppala SR, Jett AM, Alitalo A, Meri S, Akins DR (2005) Complement regulator-acquiring surface protein 1 imparts resistance to human serum in *Borrelia burgdorferi*. J Immunol 175:3299–3308

Burgdorfer W, Gage KL (1986) Susceptibility of the black-legged tick, *Ixodes scapularis*, to the Lyme disease spirochete, *Borrelia burgdorferi*. Zentralbl Bakteriol Mikrobiol Hyg [A] 263:15–20

Burgdorfer W, Barbour AG, Hayes SF, Benach JL, Grunwaldt E, Davis JP (1982) Lyme disease-a tick-borne spirochetosis? Science 216:1317–1319

Burman N, Bergstrom S, Restrepo BI, Barbour AG (1990) The variable antigens Vmp7 and Vmp21 of the relapsing fever bacterium *Borrelia hermsii* are structurally analogous to the VSG proteins of the African trypanosome. Mol Microbiol 4:1715–1726

Burroughs AL, Holdenried R (1944) Recovery of relapsing fever spirochetes from *Ornithodoros turicata* (Duges), 1876, in California. J Bacteriol 48:609

Cadavid D, Thomas DD, Crawley R, Barbour AG (1994) Variability of a bacterial surface protein and disease expression in a possible mouse model of systemic Lyme borreliosis. J Exp Med 179:631–642

Cadavid D, Pachner AR, Estanislao L, Patalapati R, Barbour AG (2001) Isogenic serotypes of *Borrelia turicatae* show different localization in the brain and skin of mice. Infect Immun 69:3389–3397

Cadavid D, Sondey M, Garcia E, Lawson CL (2006) Residual brain infection in relapsing-fever borreliosis. J Infect Dis 193:1451–1458

Calabi O (1959) The presence of plasma inhibitors during the crisis phenomenon in experimental relapsing fever (*Borrelia novyi*). J Exp Med 110:811–825

Callister SM, Schell RF, Lovrich SD (1991) Lyme disease assay which detects killed *Borrelia burgdorferi*. J Clin Microbiol 29:1773–1776

Callister SM, Schell RF, Case KL, Lovrich SD, Day SP (1993) Characterization of the borreliacidal antibody response to *Borrelia burgdorferi* in humans: a serodiagnostic test. J Infect Dis 167:158–164

Callister SM, Schell RF, Lim LC, Jobe DA, Case KL, Bryant GL, Molling PE (1994) Detection of borreliacidal antibodies by flow cytometry. An accurate, highly specific serodiagnostic test for Lyme disease. Arch Intern Med 154:1625–1632

Callister SM, Jobe DA, Schell RF, Pavia CS, Lovrich SD (1996) Sensitivity and specificity of the borreliacidal-antibody test during early Lyme disease: a "gold standard"? Clin Diagn Lab Immunol 3:399–402

Canica MM, Nato F, du Merle L, Mazie JC, Baranton G, Postic D (1993) Monoclonal antibodies for identification of *Borrelia afzelii* sp. nov. associated with late cutaneous manifestations of Lyme borreliosis. Scand J Infect Dis 25:441–448

Carter CJ, Bergstrom S, Norris SJ, Barbour AG (1994) A family of surface-exposed proteins of 20 kilodaltons in the genus *Borrelia*. Infect Immun 62:2792–2799

Chen J, Field JA, Glickstein L, Molloy PJ, Huber BT, Steere AC (1999) Association of antibiotic treatment-resistant Lyme arthritis with T cell responses to dominant epitopes of outer surface protein A of *Borrelia burgdorferi*. Arthritis Rheum 42:1813–1822

Coffey EM, Eveland WC (1967a) Experimental relapsing fever initiated by *Borella hermsi*. I. Identification of major serotypes by immunofluorescence. J Infect Dis 117:23–28

Coffey EM, Eveland WC (1967b) Experimental relapsing fever initiated by *Borrelia hermsi*. II. Sequential appearance of major serotypes in the rat. J Infect Dis 117:29–34

Coleman JL, Rogers RC, Benach JL (1992) Selection of an escape variant of *Borrelia burgdorferi* by use of bactericidal monoclonal antibodies to OspB. Infect Immun 60:3098–3104

Coleman JL, Rogers RC, Rosa PA, Benach JL (1994) Variations in the ospB gene of *Borrelia burgdorferi* result in differences in monoclonal antibody reactivity and in production of escape variants. Infect Immun 62:303–307

Connolly SE, Benach JL (2001) Cutting edge: the spirochetemia of murine relapsing fever is cleared by complement-independent bactericidal antibodies. J Immunol 167:3029–3032

Connolly SE, Benach JL (2005) The versatile roles of antibodies in *Borrelia* infections. Nat Rev Microbiol 3:411–420

Connolly SE, Thanassi DG, Benach JL (2004) Generation of a complement-independent bactericidal IgM against a relapsing fever *Borrelia*. J Immunol 172:1191–1197

Cordes FS, Roversi P, Kraiczy P, Simon MM, Brade V, Jahraus O, Wallis R, Skerka C, Zipfel PF, Wallich R, Lea SM (2005) A novel fold for the factor H-binding protein BbCRASP-1 of *Borrelia burgdorferi*. Nat Struct Mol Biol 12:276–277

Cordes FS, Kraiczy P, Roversi P, Simon MM, Brade V, Jahraus O, Wallis R, Goodstadt L, Ponting CP, Skerka C, Zipfel PF, Wallich R, Lea SM (2006) Structure-function mapping of BbCRASP-1, the key complement factor H and FHL-1 binding protein of *Borrelia burgdorferi*. Int J Med Microbiol 296 [Suppl 40]:177–184

Creson JR, Lim LC, Glowacki NJ, Callister SM, Schell RF (1996) Detection of anti-*Borrelia burgdorferi* antibody responses with the borreliacidal antibody test, indirect fluorescent-antibody assay performed by flow cytometry, and Western immunoblotting. Clin Diagn Lab Immunol 3:184–190

Croke CL, Munson EL, Lovrich SD, Christopherson JA, Remington MC, England DM, Callister SM, Schell RF (2000) Occurrence of severe destructive lyme arthritis in hamsters vaccinated with outer surface protein A and challenged with *Borrelia burgdorferi*. Infect Immun 68:658–663

Crowley H, Alroy J, Sproule TJ, Roopenian D, Huber BT (2006) The MHC class I-related FcRn ameliorates murine Lyme arthritis. Int Immunol 18:409–414

Cutler SJ, Moss J, Fukunaga M, Wright DJ, Fekade D, Warrell D (1997) Borrelia recurrentis characterization and comparison with relapsing-fever, Lyme-associated, and other *Borrelia* spp. Int J Syst Bacteriol 47:958–968

Dai Q, Restrepo BI, Porcella SF, Raffel SJ, Schwan TG, Barbour AG (2006) Antigenic variation by *Borrelia hermsii* occurs through recombination between extragenic repetitive elements on linear plasmids. Mol Microbiol 60:1329–1343

Dame TM, Orenzoff BL, Palmer LE, Furie MB (2007) IFN-{gamma} alters the response of borrelia burgdorferi-activated endothelium to favor chronic inflammation. J Immunol 178:1172–1179

Damman CJ, Eggers CH, Samuels DS, Oliver DB (2000) Characterization of *Borrelia burgdorferi* BlyA and BlyB proteins: a prophage-encoded holin-like system. J Bacteriol 182:6791–6797

Davidson MM, Chisholm SM, Wiseman AD, Joss AW, Ho-Yen DO (1996) Improved serodiagnosis of Lyme disease. Clin Mol Pathol 49:M80–M84

Davis GE (1939) Relapsing fever: *Ornithodoros hermsi* a vector in Colorado. Public Health Report 54:2178–2180

Davis GE (1940) Ticks and relapsing fever in the United States. Public Health Report 55:2347–2351

de Silva AM, Telford SR 3rd, Brunet LR, Barthold SW, Fikrig E (1996) *Borrelia burgdorferi* OspA is an arthropod-specific transmission-blocking Lyme disease vaccine. J Exp Med 183:271–275

de Silva AM, Zeidner NS, Zhang Y, Dolan MC, Piesman J, Fikrig E (1999) Influence of outer surface protein A antibody on *Borrelia burgdorferi* within feeding ticks. Infect Immun 67:30–35

Dressler F, Whalen JA, Reinhardt BN, Steere AC (1993) Western blotting in the serodiagnosis of Lyme disease. J Infect Dis 167:392–400

DuChateau BK, England DM, Callister SM, Lim LC, Lovrich SD, Schell RF (1996) Macrophages exposed to *Borrelia burgdorferi* induce Lyme arthritis in hamsters. Infect Immun 64:2540–2547

DuChateau BK, Jensen JR, England DM, Callister SM, Lovrich SD, Schell RF (1997) Macrophages and enriched populations of T lymphocytes interact synergistically for the induction of severe, destructive Lyme arthritis. Infect Immun 65:2829–2836

DuChateau BK, Munson EL, England DM, Lovrich SD, Callister SM, Jensen JR, Schell RF (1999) Macrophages interact with enriched populations of distinct T lymphocyte subsets for the induction of severe destructive Lyme arthritis. J Leukoc Biol 65:162–170

Dunne M, al-Ramadi BK, Barthold SW, Flavell RA, Fikrig E (1995) Oral vaccination with an attenuated *Salmonella typhimurium* strain expressing *Borrelia burgdorferi* OspA prevents murine Lyme borreliosis. Infect Immun 63:1611–1614

Duray PH (1989) Clinical pathologic correlations of Lyme disease. Rev Infect Dis 11 [Suppl 6]: S1487–S1493

Embers ME, Wormser GP, Schwartz I, Martin DS, Philipp MT (2007) *Borrelia burgdorferi* spirochetes that harbor only a portion of the lp28–1 plasmid elicit antibody responses detectable with the C6 test for Lyme disease. Clin Vaccine Immunol 14:90–93

Engstrom SM, Shoop E, Johnson RC (1995) Immunoblot interpretation criteria for serodiagnosis of early Lyme disease. J Clin Microbiol 33:419–427

Escudero R, Halluska ML, Backenson PB, Coleman JL, Benach JL (1997) Characterization of the physiological requirements for the bactericidal effects of a monoclonal antibody to OspB of *Borrelia burgdorferi* by confocal microscopy. Infect Immun 65:1908–1915

Exner MM, Wu X, Blanco DR, Miller JN, Lovett MA (2000) Protection elicited by native outer membrane protein Oms66 (p66) against host-adapted *Borrelia burgdorferi*: conformational nature of bactericidal epitopes. Infect Immun 68:2647–2654

Fikrig E, Barthold SW, Kantor FS, Flavell RA (1990) Protection of mice against the Lyme disease agent by immunizing with recombinant OspA. Science 250:553–536

Fikrig E, Barthold SW, Kantor FS, Flavell RA (1991) Protection of mice from Lyme borreliosis by oral vaccination with *Escherichia coli* expressing OspA. J Infect Dis 164:1224–1227

Fikrig E, Barthold SW, Kantor FS, Flavell RA (1992a) Long-term protection of mice from Lyme disease by vaccination with OspA. Infect Immun 60:773–777

Fikrig E, Telford SR 3rd, Barthold SW, Kantor FS, Spielman A, Flavell RA (1992b) Elimination of *Borrelia burgdorferi* from vector ticks feeding on OspA-immunized mice. Proc Natl Acad Sci U S A 89:5418–5421

Fikrig E, Berland R, Chen M, Williams S, Sigal LH, Flavell RA (1993) Serologic response to the *Borrelia burgdorferi* flagellin demonstrates an epitope common to a neuroblastoma cell line. Proc Natl Acad Sci U S A 90:183–187

Fikrig E, Bockenstedt LK, Barthold SW, Chen M, Tao H, Ali-Salaam P, Telford SR, Flavell RA (1994) Sera from patients with chronic Lyme disease protect mice from Lyme borreliosis. J Infect Dis 169:568–574

Fikrig E, Barthold SW, Chen M, Grewal IS, Craft J, Flavell RA (1996) Protective antibodies in murine Lyme disease arise independently of CD40 ligand. J Immunol 157:1–3

Fikrig E, Barthold SW, Chen M, Chang CH, Flavell RA (1997) Protective antibodies develop, and murine Lyme arthritis regresses, in the absence of MHC class II and CD4+ T cells. J Immunol 159:5682–5686

Filgueira L, Nestle FO, Rittig M, Joller HI, Groscurth P (1996) Human dendritic cells phagocytose and process *Borrelia burgdorferi*. J Immunol 157:2998–3005

Fraser CM, Casjens S, Huang WM, Sutton GG, Clayton R, Lathigra R, White O, Ketchum KA, Dodson R, Hickey EK, Gwinn M, Dougherty B, Tomb JF, Fleischmann RD, Richardson D, Peterson J, Kerlavage AR, Quackenbush J, Salzberg S, Hanson M, van Vugt R, Palmer N, Adams MD, Gocayne J, Weidman J, Utterback T, Watthey L, McDonald L, Artiach P, Bowman C, Garland S, Fuji C, Cotton MD, Horst K, Roberts K, Hatch B, Smith HO, Venter JC (1997) Genomic sequence of a Lyme disease spirochaete, *Borrelia burgdorferi*. Nature 390:580–586

Frey AB, Rao TD (1995) Single exposure of mice to *Borrelia burgdorferi* elicits immunoglobulin G antibodies characteristic of secondary immune response without production of interleukin-4 by immune T cells. Infect Immun 63:2596–2603

Galli G, Nuti S, Tavarini S, Galli-Stampino L, De Lalla C, Casorati G, Dellabona P, Abrignani S (2003) Innate immune responses support adaptive immunity: NKT cells induce B cell activation. Vaccine 21 [Suppl 2]:S48–S54

Garcia-Monco JC, Benach JL (1995) Lyme neuroborreliosis. Ann Neurol 37:691–702

Garcia-Monco JC, Benach JL (1997) Mechanisms of injury in Lyme neuroborreliosis. Semin Neurol 17:57–62

Garcia Monco JC, Wheeler CM, Benach JL, Furie RA, Lukehart SA, Stanek G, Steere AC (1993) Reactivity of neuroborreliosis patients (Lyme disease) to cardiolipin and gangliosides. J Neurol Sci 117:206–214

Garcia-Monco JC, Seidman RJ, Benach JL (1995) Experimental immunization with *Borrelia burgdorferi* induces development of antibodies to gangliosides. Infect Immun 63:4130–4137

Garcia-Monco JC, Miller NS, Backenson PB, Anda P, Benach JL (1997) A mouse model of Borrelia meningitis after intradermal injection. J Infect Dis 175:1243–1245

Gergel EI, Furie MB (2001) Activation of endothelium by *Borrelia burgdorferi* in vitro enhances transmigration of specific subsets of T lymphocytes. Infect Immun 69:2190–2197

Gergel EI, Furie MB (2004) Populations of human T lymphocytes that traverse the vascular endothelium stimulated by *Borrelia burgdorferi* are enriched with cells that secrete gamma interferon. Infect Immun 72:1530–1536

Ghosh S, Steere AC, Stollar BD, Huber BT (2005) In situ diversification of the antibody repertoire in chronic Lyme arthritis synovium. J Immunol 174:2860–2869

Ghosh S, Seward R, Costello CE, Stollar BD, Huber BT (2006) Autoantibodies from synovial lesions in chronic, antibiotic treatment-resistant Lyme arthritis bind cytokeratin-10. J Immunol 177:2486–2494

Gipson CL, de Silva AM (2005) Interactions of OspA monoclonal antibody C3.78 with *Borrelia burgdorferi* within ticks. Infect Immun 73:1644–1647

Golightly MG (1997) Lyme borreliosis: laboratory considerations. Semin Neurol 17:11–17

Gross DM, Forsthuber T, Tary-Lehmann M, Etling C, Ito K, Nagy ZA, Field JA, Steere AC, Huber BT (1998) Identification of LFA-1 as a candidate autoantigen in treatment-resistant Lyme arthritis. Science 281:703–706

Guerau-de-Arellano M, Huber BT (2002) Development of autoimmunity in Lyme arthritis. Curr Opin Rheumatol 14:388–393

Guinamard R, Okigaki M, Schlessinger J, Ravetch JV (2000) Absence of marginal zone B cells in Pyk-2-deficient mice defines their role in the humoral response. Nat Immunol 1:31–36

Haas KM, Poe JC, Steeber DA, Tedder TF (2005) B-1a and B-1b cells exhibit distinct developmental requirements and have unique functional roles in innate and adaptive immunity to *S. pneumoniae*. Immunity 23:7–18

Halperin JJ (2003) Lyme disease and the peripheral nervous system. Muscle Nerve 28:133–143

Halperin JJ (2005) Central nervous system Lyme disease. Curr Neurol Neurosci Rep 5:446–452

Halperin JJ, Golightly M (1992) Lyme borreliosis in Bell's palsy. Long Island Neuroborreliosis Collaborative Study Group. Neurology 42:1268–1270

Halperin JJ, Little BW, Coyle PK, Dattwyler RJ (1987) Lyme disease: cause of a treatable peripheral neuropathy. Neurology 37:1700–1706

Halperin JJ, Luft BJ, Anand AK, Roque CT, Alvarez O, Volkman DJ, Dattwyler RJ (1989) Lyme neuroborreliosis: central nervous system manifestations. Neurology 39:753–759

Halperin JJ, Volkman DJ, Wu P (1991) Central nervous system abnormalities in Lyme neuroborreliosis. Neurology 41:1571–1582

Hannier S, Liversidge J, Sternberg JM, Bowman AS (2004) Characterization of the B-cell inhibitory protein factor in *Ixodes ricinus* tick saliva: a potential role in enhanced *Borrelia burgdorferi* transmission. Immunology 113:401–408

Hansen K, Cruz M, Link H (1990) Oligoclonal *Borrelia burgdorferi*-specific IgG antibodies in cerebrospinal fluid in Lyme neuroborreliosis. J Infect Dis 161:1194–1202

Hardin JA, Steere AC, Malawista SE (1979a) Immune complexes and the evolution of Lyme arthritis. Dissemination and localization of abnormal C1q binding activity. N Engl J Med 301:1358–1363

Hardin JA, Walker LC, Steere AC, Trumble TC, Tung KS, Williams RC Jr, Ruddy S, Malawista SE (1979b) Circulating immune complexes in Lyme arthritis. Detection by the 125I-C1q binding, C1q solid phase, and Raji cell assays. J Clin Invest 63:468–477

Hardin JA, Steere AC, Malawista SE (1984) The pathogenesis of arthritis in Lyme disease: humoral immune responses and the role of intra-articular immune complexes. Yale J Biol Med 57:589–593

Hartmann K, Corvey C, Skerka C, Kirschfink M, Karas M, Brade V, Miller JC, Stevenson B, Wallich R, Zipfel PF, Kraiczy P (2006) Functional characterization of BbCRASP-2, a distinct outer membrane protein of *Borrelia burgdorferi* that binds host complement regulators factor H and FHL-1. Mol Microbiol 61:1220–1236

Hayes SF, Burgdorfer W, Barbour AG (1983) Bacteriophage in the *Ixodes dammini* spirochete, etiological agent of Lyme disease. J Bacteriol 154:1436–1439

Hellwage J, Meri T, Heikkila T, Alitalo A, Panelius J, Lahdenne P, Seppala IJ, Meri S (2001) The complement regulator factor H binds to the surface protein OspE of *Borrelia burgdorferi*. J Biol Chem 276:8427–8435

Hodzic E, Tunev S, Feng S, Freet KJ, Barthold SW (2005) Immunoglobulin-regulated expression of *Borrelia burgdorferi* outer surface protein A in vivo. Infect Immun 73:3313–3321

Holliger P, Hudson PJ (2005) Engineered antibody fragments and the rise of single domains. Nat Biotechnol 23:1126–1136

Honarvar N, Schaible UE, Galanos C, Wallich R, Simon MM (1994) A 14,000 MW lipoprotein and a glycolipid-like structure of *Borrelia burgdorferi* induce proliferation and immunoglobulin production in mouse B cells at high frequencies. Immunology 82:389–396

Horwitz MA (1984) Phagocytosis of the Legionnaires' disease bacterium (*Legionella pneumophila*) occurs by a novel mechanism: engulfment within a pseudopod coil. Cell 36:27–33

Hovis KM, McDowell JV, Griffin L, Marconi RT (2004) Identification and characterization of a linear-plasmid-encoded factor H-binding protein (FhbA) of the relapsing fever spirochete *Borrelia hermsii*. J Bacteriol 186:2612–2618

Hovis KM, Jones JP, Sadlon T, Raval G, Gordon DL, Marconi RT (2006a) Molecular analyses of the interaction of *Borrelia hermsii* FhbA with the complement regulatory proteins factor H and factor H-like protein 1. Infect Immun 74:2007–2014

Hovis KM, Schriefer ME, Bahlani S, Marconi RT (2006b) Immunological and molecular analyses of the *Borrelia hermsii* factor H and factor H-like protein 1 binding protein, FhbA: demonstration of its utility as a diagnostic marker and epidemiological tool for tick-borne relapsing fever. Infect Immun 74:4519–4529

Hovis KM, Tran E, Sundy CM, Buckles E, McDowell JV, Marconi RT (2006c) Selective binding of *Borrelia burgdorferi* OspE paralogs to factor H and serum proteins from diverse animals: possible expansion of the role of OspE in Lyme disease pathogenesis. Infect Immun 74:1967–1972

Howe TR, Mayer LW, Barbour AG (1985) A single recombinant plasmid expressing two major outer surface proteins of the Lyme disease spirochete. Science 227:645–646

Huston JS, Levinson D, Mudgett-Hunter M, Tai MS, Novotny J, Margolies MN, Ridge RJ, Bruccoleri RE, Haber E, Crea R et al (1988) Protein engineering of antibody binding sites: recovery of specific activity in an anti-digoxin single-chain Fv analogue produced in *Escherichia coli*. Proc Natl Acad Sci U S A 85:5879–5883

Isakson PC, Pure E, Vitetta ES, Krammer PH (1982) T cell-derived B cell differentiation factor(s). Effect on the isotype switch of murine B cells. J Exp Med 155:734–748

Janeway C, Travers P, Walport M, Shlomchik M (2001) Immunobiology: the immune system in health and disease. Garland Publishing, New York

Jansson C, Carlsson SA, Granlund H, Wahlberg P, Nyman D (2005) Analysis of *Borrelia burgdorferi* IgG antibodies with a combination of IgG ELISA and VlsE C6 peptide ELISA. Clin Microbiol Infect 11:147–150

Jensen JR, Du Chateau BK, Munson EL, Callister SM, Schell RF (1998) Inhibition of the production of anti-OspA borreliacidal antibody with T cells from hamsters vaccinated against *Borrelia burgdorferi*. Infect Immun 66:1507–1512

Johnson BJ, Robbins KE, Bailey RE, Cao BL, Sviat SL, Craven RB, Mayer LW, Dennis DT (1996) Serodiagnosis of Lyme disease: accuracy of a two-step approach using a flagella-based ELISA and immunoblotting. J Infect Dis 174:346–353

Johnson RC (1977) The spirochetes. Annu Rev Microbiol 31:89–106

Johnson RC, Kodner C, Russell M (1986) Passive immunization of hamsters against experimental infection with the Lyme disease spirochete. Infect Immun 53:713–714

Kaiser R (1995) Intrathecal immune response in neuroborreliosis: importance of cross-reactive antibodies. Zentralbl Bakteriol 282:303–314

Kaiser R, Rauer S (1998) Analysis of the intrathecal immune response in neuroborreliosis to a sonicate antigen and three recombinant antigens of *Borrelia burgdorferi* sensu stricto. Eur J Clin Microbiol Infect Dis 17:159–166

Kalish RA, Leong JM, Steere AC (1993) Association of treatment-resistant chronic Lyme arthritis with HLA-DR4 and antibody reactivity to OspA and OspB of *Borrelia burgdorferi*. Infect Immun 61:2774–2779

Kang I, Barthold SW, Persing DH, Bockenstedt LK (1997) T-helper-cell cytokines in the early evolution of murine Lyme arthritis. Infect Immun 65:3107–3111

Karlsson M, Mollegard I, Stiernstedt G, Wretlind B (1989) Comparison of Western blot and enzyme-linked immunosorbent assay for diagnosis of Lyme borreliosis. Eur J Clin Microbiol Infect Dis 8:871–877

Katona LI, Ayalew S, Coleman JL, Benach JL (2000) A bactericidal monoclonal antibody elicits a change in its antigen, OspB of *Borrelia burgdorferi*, that can be detected by limited proteolysis. J Immunol 164:1425–1431

Kawabata H, Masuzawa T, Yanagihara Y (1993) Genomic analysis of *Borrelia japonica* sp. nov. isolated from *Ixodes ovatus* in Japan. Microbiol Immunol 37:843–848

Keane-Myers A, Nickell SP (1995a) Role of IL-4 and IFN-gamma in modulation of immunity to *Borrelia burgdorferi* in mice. J Immunol 155:2020–2028

Keane-Myers A, Nickell SP (1995b) T cell subset-dependent modulation of immunity to *Borrelia burgdorferi* in mice. J Immunol 154:1770–1776

Keane-Myers A, Maliszewski CR, Finkelman FD, Nickell SP (1996) Recombinant IL-4 treatment augments resistance to *Borrelia burgdorferi* infections in both normal susceptible and antibody-deficient susceptible mice. J Immunol 156:2488–2494

Kochi SK, Johnson RC (1988) Role of immunoglobulin G in killing of *Borrelia burgdorferi* by the classical complement pathway. Infect Immun 56:314–321

Kochi SK, Johnson RC, Dalmasso AP (1991) Complement-mediated killing of the Lyme disease spirochete *Borrelia burgdorferi*. Role of antibody in formation of an effective membrane attack complex. J Immunol 146:3964–3970

Kochi SK, Johnson RC, Dalmasso AP (1993) Facilitation of complement-dependent killing of the Lyme disease spirochete, *Borrelia burgdorferi*, by specific immunoglobulin G Fab antibody fragments. Infect Immun 61:2532–2536

Koide S, Yang X, Huang X, Dunn JJ, Luft BJ (2005) Structure-based design of a second-generation Lyme disease vaccine based on a C-terminal fragment of *Borrelia burgdorferi* OspA. J Mol Biol 350:290–299

Kraiczy P, Skerka C, Brade V, Zipfel PF (2001a) Further characterization of complement regulator-acquiring surface proteins of *Borrelia burgdorferi*. Infect Immun 69:7800–7809

Kraiczy P, Skerka C, Kirschfink M, Brade V, Zipfel PF (2001b) Immune evasion of *Borrelia burgdorferi* by acquisition of human complement regulators FHL-1/reconectin and Factor H. Eur J Immunol 31:1674–1684

Kraiczy P, Skerka C, Kirschfink M, Zipfel PF, Brade V (2001c) Mechanism of complement resistance of pathogenic *Borrelia burgdorferi* isolates. Int Immunopharmacol 1:393–401

Kraiczy P, Hellwage J, Skerka C, Kirschfink M, Brade V, Zipfel PF, Wallich R (2003) Immune evasion of *Borrelia burgdorferi*: mapping of a complement-inhibitor factor H-binding site of BbCRASP-3, a novel member of the Erp protein family. Eur J Immunol 33:697–707

Kraiczy P, Hellwage J, Skerka C, Becker H, Kirschfink M, Simon MM, Brade V, Zipfel PF, Wallich R (2004) Complement resistance of *Borrelia burgdorferi* correlates with the expression of BbCRASP-1, a novel linear plasmid-encoded surface protein that interacts with human factor H and FHL-1 and is unrelated to Erp proteins. J Biol Chem 279:2421–2429

Kumararatne DS, MacLennan IC (1981) Cells of the marginal zone of the spleen are lymphocytes derived from recirculating precursors. Eur J Immunol 11:865–869

Kumararatne DS, MacLennan IC (1982) The origin of marginal-zone cells. Adv Exp Med Biol 149:83–90

Kumararatne DS, Bazin H, MacLennan IC (1981) Marginal zones: the major B cell compartment of rat spleens. Eur J Immunol 11:858–864

Lawrenz MB, Hardham JM, Owens RT, Nowakowski J, Steere AC, Wormser GP, Norris SJ (1999) Human antibody responses to VlsE antigenic variation protein of *Borrelia burgdorferi*. J Clin Microbiol 37:3997–4004

Le Fleche A, Postic D, Girardet K, Peter O, Baranton G (1997) Characterization of *Borrelia lusitaniae* sp. nov. by 16S ribosomal DNA sequence analysis. Int J Syst Bacteriol 47:921–925

Ledue TB, Collins MF, Craig WY (1996) New laboratory guidelines for serologic diagnosis of Lyme disease: evaluation of the two-test protocol. J Clin Microbiol 34:2343–2350

Li L, Narayan K, Pak E, Pachner AR (2006) Intrathecal antibody production in a mouse model of Lyme neuroborreliosis. J Neuroimmunol 173:56–68

Liang FT, Philipp MT (1999) Analysis of antibody response to invariable regions of VlsE, the variable surface antigen of *Borrelia burgdorferi*. Infect Immun 67:6702–6706

Liang FT, Alvarez AL, Gu Y, Nowling JM, Ramamoorthy R, Philipp MT (1999a) An immunodominant conserved region within the variable domain of VlsE, the variable surface antigen of *Borrelia burgdorferi*. J Immunol 163:5566–5573

Liang FT, Steere AC, Marques AR, Johnson BJ, Miller JN, Philipp MT (1999b) Sensitive and specific serodiagnosis of Lyme disease by enzyme-linked immunosorbent assay with a peptide based on an immunodominant conserved region of *Borrelia burgdorferi* vlsE. J Clin Microbiol 37:3990–3996

Liang FT, Aberer E, Cinco M, Gern L, Hu CM, Lobet YN, Ruscio M, Voet PE Jr, Weynants VE, Philipp MT (2000a) Antigenic conservation of an immunodominant invariable region of the VlsE lipoprotein among European pathogenic genospecies of *Borrelia burgdorferi* SL. J Infect Dis 182:1455–1462

Liang FT, Jacobson RH, Straubinger RK, Grooters A, Philipp MT (2000b) Characterization of a *Borrelia burgdorferi* VlsE invariable region useful in canine Lyme disease serodiagnosis by enzyme-linked immunosorbent assay. J Clin Microbiol 38:4160–4166

Liang FT, Jacobs MB, Bowers LC, Philipp MT (2002) An immune evasion mechanism for spirochetal persistence in Lyme borreliosis. J Exp Med 195:415–422

Liang FT, Yan J, Mbow ML, Sviat SL, Gilmore RD, Mamula M, Fikrig E (2004) *Borrelia burgdorferi* changes its surface antigenic expression in response to host immune responses. Infect Immun 72:5759–5767

Linder S, Heimerl C, Fingerle V, Aepfelbacher M, Wilske B (2001) Coiling phagocytosis of *Borrelia burgdorferi* by primary human macrophages is controlled by CDC42Hs and Rac1 and involves recruitment of Wiskott-Aldrich syndrome protein and Arp2/3 complex. Infect Immun 69:1739–1746

Lovrich SD, Callister SM, Schmitz JL, Alder JD, Schell RF (1991) Borreliacidal activity of sera from hamsters infected with the Lyme disease spirochete. Infect Immun 59:2522–2528

Lovrich SD, Jobe DA, Schell RF, Callister SM (2005) Borreliacidal OspC antibodies specific for a highly conserved epitope are immunodominant in human lyme disease and do not occur in mice or hamsters. Clin Diagn Lab Immunol 12:746–751

Luke CJ, Huebner RC, Kasmiersky V, Barbour AG (1997) Oral delivery of purified lipoprotein OspA protects mice from systemic infection with *Borrelia burgdorferi*. Vaccine 15:739–746

Ma Y, Weis JJ (1993) Borrelia burgdorferi outer surface lipoproteins OspA and OspB possess B-cell mitogenic and cytokine-stimulatory properties. Infect Immun 61:3843–3853

Ma J, Gingrich-Baker C, Franchi PM, Bulger P, Coughlin RT (1995) Molecular analysis of neutralizing epitopes on outer surface proteins A and B of *Borrelia burgdorferi*. Infect Immun 63:2221–2227

Magnarelli LA, Lawrenz M, Norris SJ, Fikrig E (2002) Comparative reactivity of human sera to recombinant VlsE and other *Borrelia burgdorferi* antigens in class-specific enzyme-linked immunosorbent assays for Lyme borreliosis. J Med Microbiol 51:649–655

Martin F, Kearney JF (2001) B1 cells: similarities and differences with other B cell subsets. Curr Opin Immunol 13:195–201

Martin R, Martens U, Sticht-Groh V, Dorries R, Kruger H (1988) Persistent intrathecal secretion of oligoclonal, *Borrelia burgdorferi*-specific IgG in chronic meningoradiculomyelitis. J Neurol 235:229–233

Matyniak JE, Reiner SL (1995) T helper phenotype and genetic susceptibility in experimental Lyme disease. J Exp Med 181:1251–1254

Mbow ML, Gilmore RD Jr, Titus RG (1999) An OspC-specific monoclonal antibody passively protects mice from tick-transmitted infection by *Borrelia burgdorferi* B31. Infect Immun 67:5470–5472

McDowell JV, Tran E, Hamilton D, Wolfgang J, Miller K, Marconi RT (2003) Analysis of the ability of spirochete species associated with relapsing fever, avian borreliosis, and epizootic bovine abortion to bind factor H and cleave c3b. J Clin Microbiol 41:3905–3910

McDowell JV, Wolfgang J, Senty L, Sundy CM, Noto MJ, Marconi RT (2004) Demonstration of the involvement of outer surface protein E coiled coil structural domains and higher order structural elements in the binding of infection-induced antibody and the complement-regulatory protein, factor H. J Immunol 173:7471–7480

McDowell JV, Hovis KM, Zhang H, Tran E, Lankford J, Marconi RT (2006) Evidence that the BBA68 protein (BbCRASP-1) of the Lyme disease spirochetes does not contribute to factor H-mediated immune evasion in humans and other animals. Infect Immun 74:3030–3034

McKisic MD, Barthold SW (2000) T-cell-independent responses to *Borrelia burgdorferi* are critical for protective immunity and resolution of lyme disease. Infect Immun 68:5190–5197

Meier JT, Simon MI, Barbour AG (1985) Antigenic variation is associated with DNA rearrangements in a relapsing fever *Borrelia*. Cell 41:403–409

Meri T, Cutler SJ, Blom AM, Meri S, Jokiranta TS (2006) Relapsing fever spirochetes *Borrelia recurrentis* and *B. duttonii* acquire complement regulators C4b-binding protein and factor H. Infect Immun 74:4157–4163

Modolell M, Schaible UE, Rittig M, Simon MM (1994) Killing of *Borrelia burgdorferi* by macrophages is dependent on oxygen radicals and nitric oxide and can be enhanced by antibodies to outer surface proteins of the spirochete. Immunol Lett 40:139–146

Mogilyansky E, Loa CC, Adelson ME, Mordechai E, Tilton RC (2004) Comparison of Western immunoblotting and the C6 Lyme antibody test for laboratory detection of Lyme disease. Clin Diagn Lab Immunol 11:924–929

Montecino-Rodriguez E, Dorshkind K (2006) New perspectives in B-1 B cell development and function. Trends Immunol 27:428–433

Montgomery RR, Malawista SE (1996) Entry of *Borrelia burgdorferi* into macrophages is end-on and leads to degradation in lysosomes. Infect Immun 64:2867–2872

Montgomery RR, Nathanson MH, Malawista SE (1993) The fate of *Borrelia burgdorferi*, the agent for Lyme disease, in mouse macrophages. Destruction, survival, recovery. J Immunol 150:909–915

Montgomery RR, Nathanson MH, Malawista SE (1994) Fc- and non-Fc-mediated phagocytosis of *Borrelia burgdorferi* by macrophages. J Infect Dis 170:890–893

Montgomery RR, Lusitani D, de Boisfleury Chevance A, Malawista SE (2002) Human phagocytic cells in the early innate immune response to *Borrelia burgdorferi*. J Infect Dis 185:1773–1779

Morshed MG, Yokota M, Nakazawa T, Konishi H (1993) Transfer of antibody against *Borrelia duttonii* from mother to young in ddY mice. Infect Immun 61:4147–4152

Munson EL, Du Chateau BK, Jobe DA, Lovrich SD, Callister SM, Schell RF (2000) Production of borreliacidal antibody to outer surface protein A in vitro and modulation by interleukin-4. Infect Immun 68:5496–5501

Munson EL, Du Chateau BK, Jensen JR, Callister SM, DeCoster DJ, Schell RF (2002) Gamma interferon inhibits production of Anti-OspA borreliacidal antibody in vitro. Clin Diagn Lab Immunol 9:1095–1101

Munson EL, DeCoster DJ, Nardelli DT, England DM, Callister SM, Schell RF (2004) Neutralization of gamma interferon augments borreliacidal antibody production and severe destructive Lyme arthritis in C3H/HeJ mice. Clin Diagn Lab Immunol 11:35–41

Munson EL, Nardelli DT, Luk KH, Remington MC, Callister SM, Schell RF (2006) Interleukin-6 promotes anti-OspA borreliacidal antibody production in vitro. Clin Vaccine Immunol 13:19–25

Murray N, Kristoferitsch W, Stanek G, Steck AJ (1986) Specificity of CSF antibodies against components of *Borrelia burgdorferi* in patients with meningopolyneuritis Garin-Bujadoux-Bannwarth. J Neurol 233:224–227

Nassal M, Skamel C, Kratz PA, Wallich R, Stehle T, Simon MM (2005) A fusion product of the complete *Borrelia burgdorferi* outer surface protein A (OspA) and the hepatitis B virus capsid protein is highly immunogenic and induces protective immunity similar to that seen with an effective lipidated OspA vaccine formula. Eur J Immunol 35:655–665

Neubert U, Schaller M, Januschke E, Stolz W, Schmieger H (1993) Bacteriophages induced by ciprofloxacin in a *Borrelia burgdorferi* skin isolate. Zentralbl Bakteriol 279:307–315

Newman K Jr, Johnson RC (1981) In vivo evidence that an intact lytic complement pathway is not essential for successful removal of circulating *Borrelia turicatae* from mouse blood. Infect Immun 31:465–469

Newman K Jr, Johnson RC (1984) T-cell-independent elimination of *Borrelia turicatae*. Infect Immun 45:572–576

Nieva J, Kerwin L, Wentworth AD, Lerner RA, Wentworth P Jr (2006) Immunoglobulins can utilize riboflavin (Vitamin B2) to activate the antibody-catalyzed water oxidation pathway. Immunol Lett 103:33–38

Novy FG, Knapp RE (1906) Studies on *Spirillum obermeieri* and related organisms. J Infect Dis 3:291–393

Nowling JM, Philipp MT (1999) Killing of Borrelia burgdorferi by antibody elicited by OspA vaccine is inefficient in the absence of complement. Infect Immun 67:443–445

Ochsenbein AF, Fehr T, Lutz C, Suter M, Brombacher F, Hengartner H, Zinkernagel RM (1999) Control of early viral and bacterial distribution and disease by natural antibodies. Science 286:2156–2159

Oliver AM, Martin F, Kearney JF (1999) IgMhighCD21high lymphocytes enriched in the splenic marginal zone generate effector cells more rapidly than the bulk of follicular B cells. J Immunol 162:7198–7207

Pachner AR, Ricalton NS (1992) Western blotting in evaluating Lyme seropositivity and the utility of a gel densitometric approach. Neurology 42:2185–2192

Pennington PM, Allred CD, West CS, Alvarez R, Barbour AG (1997) Arthritis severity and spirochete burden are determined by serotype in the *Borrelia turicatae*-mouse model of Lyme disease. Infect Immun 65:285–292

Philipp MT, Bowers LC, Fawcett PT, Jacobs MB, Liang FT, Marques AR, Mitchell PD, Purcell JE, Ratterree MS, Straubinger RK (2001) Antibody response to IR6, a conserved immunodominant region of the VlsE lipoprotein, wanes rapidly after antibiotic treatment of *Borrelia burgdorferi* infection in experimental animals and in humans. J Infect Dis 184:870–878

Philipp MT, Wormser GP, Marques AR, Bittker S, Martin DS, Nowakowski J, Dally LG (2005) A decline in C6 antibody titer occurs in successfully treated patients with culture-confirmed early localized or early disseminated Lyme borreliosis. Clin Diagn Lab Immunol 12:1069–1074

Piesman J, Sinsky RJ (1988) Ability to Ixodes scapularis, *Dermacentor variabilis*, and *Amblyomma americanum* (Acari: Ixodidae) to acquire, maintain, and transmit Lyme disease spirochetes (*Borrelia burgdorferi*). J Med Entomol 25:336–339

Pillai S, Cariappa A, Moran ST (2005) Marginal zone B cells. Annu Rev Immunol 23:161–196

Plasterk RH, Simon MI, Barbour AG (1985) Transposition of structural genes to an expression sequence on a linear plasmid causes antigenic variation in the bacterium *Borrelia hermsii*. Nature 318:257–263

Porcella SF, Raffel SJ, Schrumpf ME, Schriefer ME, Dennis DT, Schwan TG (2000) Serodiagnosis of louse-borne relapsing fever with glycerophosphodiester phosphodiesterase (GlpQ) from *Borrelia recurrentis*. J Clin Microbiol 38:3561–3571

Probert WS, Crawford M, LeFebvre RB (1997) Antibodies to OspB prevent infection of C3H mice challenged with *Borrelia burgdorferi* isolates expressing truncated OspB antigens. Vaccine 15:15–19

Rao TD, Frey AB (1995) Protective resistance to experimental *Borrelia burgdorferi* infection of mice by adoptive transfer of a CD4+ T cell clone. Cell Immunol 162:225–234

Ras NM, Lascola B, Postic D, Cutler SJ, Rodhain F, Baranton G, Raoult D (1996) Phylogenesis of relapsing fever *Borrelia* spp. Int J Syst Bacteriol 46:859–865

Rathinavelu S, Broadwater A, de Silva AM (2003) Does host complement kill *Borrelia burgdorferi* within ticks? Infect Immun 71:822–829

Remington MC, Munson EL, Callister SM, Molitor ML, Christopherson JA, DeCoster DJ, Lovrich SD, Schell RF (2001) Interleukin-6 enhances production of anti-OspC immunoglobulin G2b borreliacidal antibody. Infect Immun 69:4268–4275

Rittig MG, Krause A, Haupl T, Schaible UE, Modolell M, Kramer MD, Lutjen-Drecoll E, Simon MM, Burmester GR (1992) Coiling phagocytosis is the preferential phagocytic mechanism for *Borrelia burgdorferi*. Infect Immun 60:4205–4212

Rittig MG, Jagoda JC, Wilske B, Murgia R, Cinco M, Repp R, Burmester GR, Krause A (1998) Coiling phagocytosis discriminates between different spirochetes and is enhanced by phorbol myristate acetate and granulocyte-macrophage colony-stimulating factor. Infect Immun 66:627–635

Rohrer JW, Gershon RK, Lynch RG, Kemp JD (1983) Enhancement of B lymphocyte secretory differentiation by a Ly 1+,2-,Qa-1+ helper T cell subset that sees both antigen and determinants on immunoglobulin. J Mol Cell Immunol 1:50–64

Rose CD, Fawcett PT, Gibney KM (2001) Arthritis following recombinant outer surface protein A vaccination for Lyme disease. J Rheumatol 28:2555–2557

Rossmann E, Kitiratschky V, Hofmann H, Kraiczy P, Simon MM, Wallich R (2006) BbCRASP-1 of the Lyme disease spirochetes is expressed in humans and induces antibody responses restricted to non-denatured structural determinants. Infect Immun 74:7024–7028

Rousselle JC, Callister SM, Schell RF, Lovrich SD, Jobe DA, Marks JA, Wieneke CA (1998) Borreliacidal antibody production against outer surface protein C of *Borrelia burgdorferi*. J Infect Dis 178:733–741

Russell H, Sampson JS, Schmid GP, Wilkinson HW, Plikaytis B (1984) Enzyme-linked immunosorbent assay and indirect immunofluorescence assay for Lyme disease. J Infect Dis 149:465–470

Sadziene A, Jonsson M, Bergstrom S, Bright RK, Kennedy RC, Barbour AG (1994) A bactericidal antibody to *Borrelia burgdorferi* is directed against a variable region of the OspB protein. Infect Immun 62:2037–2045

Satoskar AR, Elizondo J, Monteforte GM, Stamm LM, Bluethmann H, Katavolos P, Telford SR 3rd (2000) Interleukin-4-deficient BALB/c mice develop an enhanced Th1-like response but control cardiac inflammation following *Borrelia burgdorferi* infection. FEMS Microbiol Lett 183:319–325

Schaible UE, Kramer MD, Eichmann K, Modolell M, Museteanu C, Simon MM (1990) Monoclonal antibodies specific for the outer surface protein A (OspA) of *Borrelia burgdorferi* prevent Lyme borreliosis in severe combined immunodeficiency (SCID) mice. Proc Natl Acad Sci U S A 87:3768–3772

Schaible UE, Wallich R, Kramer MD, Nerz G, Stehle T, Museteanu C, Simon MM (1994) Protection against *Borrelia burgdorferi* infection in SCID mice is conferred by presensitized spleen cells and partially by B but not T cells alone. Int Immunol 6:671–681

Scheckelhoff MR, Telford SR, Hu LT (2006) Protective efficacy of an oral vaccine to reduce carriage of *Borrelia burgdorferi* (strain N40) in mouse and tick reservoirs. Vaccine 24:1949–1957

Schluesener HJ, Martin R, Sticht-Groh V (1989) Autoimmunity in Lyme disease: molecular cloning of antigens recognized by antibodies in the cerebrospinal fluid. Autoimmunity 2:323–330

Schmitz JL, Lovrich SD, Callister SM, Schell RF (1991) Depletion of complement and effects on passive transfer of resistance to infection with *Borrelia burgdorferi*. Infect Immun 59:3815–3818

Schmitz JL, Schell RF, Callister SM, Lovrich SD, Day SP, Coe JE (1992) Immunoglobulin G2 confers protection against *Borrelia burgdorferi* infection in LSH hamsters. Infect Immun 60:2677–2682

Schoenfeld R, Araneo B, Ma Y, Yang LM, Weis JJ (1992) Demonstration of a B-lymphocyte mitogen produced by the Lyme disease pathogen, *Borrelia burgdorferi*. Infect Immun 60:455–464

Schroder NW, Schombel U, Heine H, Gobel UB, Zahringer U, Schumann RR (2003) Acylated cholesteryl galactoside as a novel immunogenic motif in *Borrelia burgdorferi* sensu stricto. J Biol Chem 278:33645–33653

Schutzer SE, Coyle PK, Krupp LB, Deng Z, Belman AL, Dattwyler R, Luft BJ (1997) Simultaneous expression of *Borrelia* OspA and OspC and IgM response in cerebrospinal fluid in early neurologic Lyme disease. J Clin Invest 100:763–767

Schwan TG, Piesman J, Golde WT, Dolan MC, Rosa PA (1995) Induction of an outer surface protein on *Borrelia burgdorferi* during tick feeding. Proc Natl Acad Sci U S A 92:2909–2913

Schwan TG, Schrumpf ME, Hinnebusch BJ, Anderson DE Jr, Konkel ME (1996) GlpQ: an antigen for serological discrimination between relapsing fever and Lyme borreliosis. J Clin Microbiol 34:2483–2492

Schwan TG, Battisti JM, Porcella SF, Raffel SJ, Schrumpf ME, Fischer ER, Carroll JA, Stewart PE, Rosa P, Somerville GA (2003) Glycerol-3-phosphate acquisition in spirochetes: distribution and biological activity of glycerophosphodiester phosphodiesterase (GlpQ) among *Borrelia* species. J Bacteriol 185:1346–1356

Scriba M, Ebrahim JS, Schlott T, Eiffert H (1993) The 39-kilodalton protein of *Borrelia burgdorferi*: a target for bactericidal human monoclonal antibodies. Infect Immun 61:4523–4526

Sethi N, Sondey M, Bai Y, Kim KS, Cadavid D (2006) Interaction of a neurotropic strain of *Borrelia turicatae* with the cerebral microcirculation system. Infect Immun 74:6408–6418

Sigal LH (1993) Cross-reactivity between *Borrelia burgdorferi* flagellin and a human axonal 64,000 molecular weight protein. J Infect Dis 167:1372–1378

Sigal LH (1997) Lyme disease: a review of aspects of its immunology and immunopathogenesis. Annu Rev Immunol 15:63–92

Sigal LH, Tatum AH (1988) Lyme disease patients' serum contains IgM antibodies to *Borrelia burgdorferi* that cross-react with neuronal antigens. Neurology 38:1439–1442

Sigal LH, Williams S (1997) A monoclonal antibody to *Borrelia burgdorferi* flagellin modifies neuroblastoma cell neuritogenesis in vitro: a possible role for autoimmunity in the neuropathy of Lyme disease. Infect Immun 65:1722–1728

Sigal LH, Zahradnik JM, Lavin P, Patella SJ, Bryant G, Haselby R, Hilton E, Kunkel M, Adler-Klein D, Doherty T, Evans J, Molloy PJ, Seidner AL, Sabetta JR, Simon HJ, Klempner MS, Mays J, Marks D, Malawista SE (1998) A vaccine consisting of recombinant *Borrelia burgdorferi* outer-surface protein A to prevent Lyme disease. Recombinant Outer-Surface Protein A Lyme Disease Vaccine Study Consortium. N Engl J Med 339:216–222

Simon MM, Schaible UE, Kramer MD, Eckerskorn C, Museteanu C, Muller-Hermelink HK, Wallich R (1991) Recombinant outer surface protein a from *Borrelia burgdorferi* induces antibodies protective against spirochetal infection in mice. J Infect Dis 164:123–132

Skamel C, Ploss M, Bottcher B, Stehle T, Wallich R, Simon MM, Nassal M (2006) Hepatitis B virus capsid-like particles can display the complete, dimeric outer surface protein C and stimulate production of protective antibody responses against *Borrelia burgdorferi* infection. J Biol Chem 281:17474–17481

Snapper CM, Paul WE (1987) Interferon-gamma and B cell stimulatory factor-1 reciprocally regulate Ig isotype production. Science 236:944–947

Sole M, Bantar C, Indest K, Gu Y, Ramamoorthy R, Coughlin R, Philipp MT (1998) *Borrelia burgdorferi* escape mutants that survive in the presence of antiserum to the OspA vaccine are killed when complement is also present. Infect Immun 66:2540–2546

Sood SK, Rubin LG, Blader ME, Ilowite NT (1993) Positive serology for Lyme borreliosis in patients with juvenile rheumatoid arthritis in a Lyme borreliosis endemic area: analysis by immunoblot. J Rheumatol 20:739–741

Southern PM, Sanford JP (1969) Relapsing fever. A clinical and microbiological review. Medicine 48:129–149

Spagnuolo PJ, Butler T, Bloch EH, Santoro C, Tracy JW, Johnson RC (1982) Opsonic requirements for phagocytosis of *Borrelia hermsii* by human polymorphonuclear leukocytes. J Infect Dis 145:358–364

Stanek G (1991) Laboratory diagnosis and seroepidemiology of Lyme borreliosis. Infection 19:263–267

Steere AC (2001) Lyme disease. N Engl J Med 345:115–125

Steere AC, Hardin JA, Malawista SE (1977) Erythema chronicum migrans and Lyme arthritis: cryoimmunoglobulins and clinical activity of skin and joints. Science 196:1121–1122

Steere AC, Hardin JA, Ruddy S, Mummaw JG, Malawista SE (1979) Lyme arthritis: correlation of serum and cryoglobulin IgM with activity, and serum IgG with remission. Arthritis Rheum 22:471–483

Steere AC, Berardi VP, Weeks KE, Logigian EL, Ackermann R (1990) Evaluation of the intrathecal antibody response to Borrelia burgdorferi as a diagnostic test for Lyme neuroborreliosis. J Infect Dis 161:1203–1209

Steere AC, Gross D, Meyer AL, Huber BT (2001) Autoimmune mechanisms in antibiotic treatment-resistant lyme arthritis. J Autoimmun 16:263–268

Steere AC, Sikand VK, Meurice F, Parenti DL, Fikrig E, Schoen RT, Nowakowski J, Schmid CH, Laukamp S, Buscarino C, Krause DS (1998) Vaccination against Lyme disease with recombinant

Borrelia burgdorferi outer-surface lipoprotein A with adjuvant. Lyme Disease Vaccine Study Group. N Engl J Med 339:209–215

Steere AC, Falk B, Drouin EE, Baxter-Lowe LA, Hammer J, Nepom GT (2003) Binding of outer surface protein A and human lymphocyte function-associated antigen 1 peptides to HLA-DR molecules associated with antibiotic treatment-resistant Lyme arthritis. Arthritis Rheum 48:534–540

Stevens TL, Bossie A, Sanders VM, Fernandez-Botran R, Coffman RL, Mosmann TR, Vitetta ES (1988) Regulation of antibody isotype secretion by subsets of antigen-specific helper T cells. Nature 334:255–258

Stevenson B, Porcella SF, Oie KL, Fitzpatrick CA, Raffel SJ, Lubke L, Schrumpf ME, Schwan TG (2000) The relapsing fever spirochete *Borrelia hermsii* contains multiple, antigen-encoding circular plasmids that are homologous to the cp32 plasmids of Lyme disease spirochetes. Infect Immun 68:3900–3908

Stevenson B, El-Hage N, Hines MA, Miller JC, Babb K (2002) Differential binding of host complement inhibitor factor H by Borrelia burgdorferi Erp surface proteins: a possible mechanism underlying the expansive host range of Lyme disease spirochetes. Infect Immun 70:491–497

Stoenner HG, Dodd T, Larsen C (1982) Antigenic variation of *Borrelia hermsii*. J Exp Med 156:1297–1311

Tai KF, Ma Y, Weis JJ (1994) Normal human B lymphocytes and mononuclear cells respond to the mitogenic and cytokine-stimulatory activities of *Borrelia burgdorferi* and its lipoprotein OspA. Infect Immun 62:520–528

Takayama K, Rothenberg RJ, Barbour AG (1987) Absence of lipopolysaccharide in the Lyme disease spirochete, *Borrelia burgdorferi*. Infect Immun 55:2311–2313

Trollmo C, Meyer AL, Steere AC, Hafler DA, Huber BT (2001) Molecular mimicry in Lyme arthritis demonstrated at the single cell level: LFA-1 alpha L is a partial agonist for outer surface protein A-reactive T cells. J Immunol 166:5286–5291

U.S. Department of Health and Human Services CfDCaP (1995) Recommendations for test performance and interpretation from the Second National Conference on Serologic Diagnosis of Lyme Disease. Morb Mortal Wkly Rep 44:590–591

Ulbrandt ND, Cassatt DR, Patel NK, Roberts WC, Bachy CM, Fazenbaker CA, Hanson MS (2001) Conformational nature of the Borrelia burgdorferi decorin binding protein A epitopes that elicit protective antibodies. Infect Immun 69:4799–4807

van Dam AP, Oei A, Jaspars R, Fijen C, Wilske B, Spanjaard L, Dankert J (1997) Complement-mediated serum sensitivity among spirochetes that cause Lyme disease. Infect Immun 65:1228–1236

Vincent MS, Gumperz JE, Brenner MB (2003) Understanding the function of CD1-restricted T cells. Nat Immunol 4:517–23

von Lackum K, Miller JC, Bykowski T, Riley SP, Woodman ME, Brade V, Kraiczy P, Stevenson B, Wallich R (2005) Borrelia burgdorferi regulates expression of complement regulator-acquiring surface protein 1 during the mammal-tick infection cycle. Infect Immun 73:7398–7405

Wang G, van Dam AP, Le Fleche A, Postic D, Peter O, Baranton G, de Boer R, Spanjaard L, Dankert J (1997) Genetic and phenotypic analysis of *Borrelia valaisiana* sp. nov. (Borrelia genomic groups VS116 and M19). Int J Syst Bacteriol 47:926–932

Weinstein A, Britchkov M (2002) Lyme arthritis and post-Lyme disease syndrome. Curr Opin Rheumatol 14:383–387

Wentworth AD, Jones LH, Wentworth P Jr, Janda KD, Lerner RA (2000) Antibodies have the intrinsic capacity to destroy antigens. Proc Natl Acad Sci U S A 97:10930–10935

Wentworth P Jr, McDunn JE, Wentworth AD, Takeuchi C, Nieva J, Jones T, Bautista C, Ruedi JM, Gutierrez A, Janda KD, Babior BM, Eschenmoser A, Lerner RA (2002) Evidence for antibody-catalyzed ozone formation in bacterial killing and inflammation. Science 298:2195–2199

Wentworth P Jr, Wentworth AD, Zhu X, Wilson IA, Janda KD, Eschenmoser A, Lerner RA (2003) Evidence for the production of trioxygen species during antibody-catalyzed chemical modification of antigens. Proc Natl Acad Sci U S A 100:1490–1493

Wheeler CM, Garcia Monco JC, Benach JL, Golightly MG, Habicht GS, Steere AC (1993) Nonprotein antigens of *Borrelia burgdorferi*. J Infect Dis 167:665–674

Whitmire WM, Garon CF (1993) Specific and nonspecific responses of murine B cells to membrane blebs of *Borrelia burgdorferi*. Infect Immun 61:1460–1467

Wilkinson HW (1984) Immunodiagnostic tests for Lyme disease. Yale J Biol Med 57:567–572

Willett TA, Meyer AL, Brown EL, Huber BT (2004) An effective second-generation outer surface protein A-derived Lyme vaccine that eliminates a potentially autoreactive T cell epitope. Proc Natl Acad Sci U S A 101:1303–1308

Wilske B, Schierz G, Preac-Mursic V, von Busch K, Kuhbeck R, Pfister HW, Einhaupl K (1986) Intrathecal production of specific antibodies against *Borrelia burgdorferi* in patients with lymphocytic meningoradiculitis (Bannwarth's syndrome). J Infect Dis 153:304–314

Xu Q, Seemanapalli SV, McShan K, Liang FT (2006) Constitutive expression of outer surface protein C diminishes the ability of *Borrelia burgdorferi* to evade specific humoral immunity. Infect Immun 74:5177–5184

Yang JQ, Singh AK, Wilson MT, Satoh M, Stanic AK, Park JJ, Hong S, Gadola SD, Mizutani A, Kakumanu SR, Reeves WH, Cerundolo V, Joyce S, Van Kaer L, Singh RR (2003) Immunoregulatory role of CD1d in the hydrocarbon oil-induced model of lupus nephritis. J Immunol 171:2142–2153

Yang L, Ma Y, Schoenfeld R, Griffiths M, Eichwald E, Araneo B, Weis JJ (1992) Evidence for B-lymphocyte mitogen activity in *Borrelia burgdorferi*-infected mice. Infect Immun 60:3033–3041

Yokota M, Morshed MG, Nakazawa T, Konishi H (1997) Protective activity of *Borrelia duttonii*-specific immunoglobulin subclasses in mice. J Med Microbiol 46:675–680

Yu D, Liang J, Yu H, Wu H, Xu C, Liu J, Lai R (2006) A tick B-cell inhibitory protein from salivary glands of the hard tick, *Hyalomma asiaticum asiaticum*. Biochem Biophys Res Commun 343:585–590

Yu Z, Tu J, Chu YH (1997) Confirmation of cross-reactivity between Lyme antibody H9724 and human heat shock protein 60 by a combinatorial approach. Anal Chem 69:4515–4518

Zhang JR, Hardham JM, Barbour AG, Norris SJ (1997) Antigenic variation in Lyme disease borreliae by promiscuous recombination of VMP-like sequence cassettes. Cell 89:275–285

Zhong W, Stehle T, Museteanu C, Siebers A, Gern L, Kramer M, Wallich R, Simon MM (1997) Therapeutic passive vaccination against chronic Lyme disease in mice. Proc Natl Acad Sci U S A 94:12533–12538

Zhong W, Gern L, Stehle T, Museteanu C, Kramer M, Wallich R, Simon MM (1999) Resolution of experimental and tick-borne *Borrelia burgdorferi* infection in mice by passive, but not active immunization using recombinant OspC. Eur J Immunol 29:946–957

Zhu X, Wentworth P Jr, Wentworth AD, Eschenmoser A, Lerner RA, Wilson IA (2004) Probing the antibody-catalyzed water-oxidation pathway at atomic resolution. Proc Natl Acad Sci U S A 101:2247–2252

A Distinct Role for B1b Lymphocytes in T Cell-Independent Immunity

K. R. Alugupalli

1	Introduction	106
2	T Cell-Dependent and -Independent Antigens	107
3	B Cell Subsets	107
4	B1b Cells	108
5	Role of B1b Cells in T Cell-Independent Responses	109
5.1	Immunity to *Borrelia hermsii*	109
5.2	Immunity to *Streptococcus pneumoniae*	113
5.3	Response to NP-Ficoll, a Model TI-2 Antigen	114
6	Activation of Antigen-Specific T Cell-Independent B Cell Responses	115
6.1	Role of Btk in T Cell-Independent B Cell Responses	115
6.2	Role of Co-stimulatory Signals in T Cell-Independent B Cell Responses	116
7	Impaired T Cell-Independent B Cell Responses	120
8	Memory B1b Cells	121
9	Concluding Remarks	123
	References	124

Abstract Pathogenesis of infectious disease is not only determined by the virulence of the microbe but also by the immune status of the host. Vaccination is the most effective means to control infectious diseases. A hallmark of the adaptive immune system is the generation of B cell memory, which provides a long-lasting protective antibody response that is central to the concept of vaccination. Recent studies revealed a distinct function for B1b lymphocytes, a minor subset of mature B cells that closely resembles that of memory B cells in a number of aspects. In contrast to the development of conventional B cell memory, which requires the formation of germinal centers and T cells, the development of B1b cell-mediated long-lasting antibody responses occurs independent of T cell help. T cell-independent (TI) antigens are important virulence factors expressed by a number of bacterial pathogens, including those associated with biological

K. R. Alugupalli
Department of Microbiology and Immunology, Kimmel Cancer Center, Thomas Jefferson University, 233 South 10th Street, BLSB 726, Philadelphia, PA 19107, USA
e-mail: kishore.alugupalli@mail.jci.tju.edu

T. Manser (ed.) *Specialization and Complementation of Humoral Immune Responses to Infection. Current Topics in Microbiology and Immunology 319.*
© Springer-Verlag Berlin Heidelberg 2008

threats. TI antigens cannot be processed and presented to T cells and therefore are known to possess restricted T cell-dependent (TD) immunogenicity. Nevertheless, specific recognition of TI antigens by B1b cells and the highly protective antibody responses mounted by them clearly indicate a crucial role for this subset of B cells. Understanding the mechanisms of long-term immunity conferred by B1b cells may lead to improved vaccine efficacy for a variety of TI antigens.

Abbreviations AID: Activation-induced cytidinedeaminase; BCR: B cell antigen receptor; Btk: Bruton's tyrosine kinase; CSR: Class switch-recombination; FO: Follicular; LPS: Lipopolysaccharide; MZ: Marginal zone; NP: 4-Hydroxy-3-Nitrophenyl-Acetyl; PerC: Peritoneal cavity; PS: Polysaccharide; Rag1: Recombination-activating gene 1; SHM: Somatic hypermutation; TD: T cell-dependent; TI: T cell-independent; TI-1: T cell-independent type 1; TI-2: T cell-independent type 2; TLR: Toll-like receptor; Xid: X-linked immunodeficiency; XLA: X-linked agammaglobulinemia

1 Introduction

Infectious diseases are the leading cause of mortality and morbidity worldwide. The acquisition of virulence-encoding genetic elements by horizontal transfer, the high mutation rates of pathogens, and the emergence of antibiotic resistance make it difficult to control infectious diseases by therapeutic means. Vaccination is the most effective way to control infectious diseases because it induces long-lasting immunity. A hallmark of the adaptive immune system is the development of B cell memory, which provides a protective antibody response upon re-exposure to the same antigen and is central to the concept of vaccination (MacLennan et al. 2000). T cells are crucial for the formation of germinal centers, a specialized microenvironment of the secondary lymphoid organs in which proliferation of antigen-specific B cells, affinity maturation of B cell antigen receptor (BCR) by somatic hypermutation (SHM), and alteration of antibody isotype by class-switch recombination (CSR) occur. The ultimate outcome of the germinal center reaction is the development of memory B cells and long-lived plasma cells that generate high-affinity antibodies of different Ig isotypes (McHeyzer-Williams et al. 2001; McHeyzer-Williams 2003). T cell-independent (TI) antibody responses are highly protective and develop much faster than T cell-dependent (TD) antibody responses (Maizels and Bothwell 1985; Vos et al. 2000; Martin and Kearney 2001). In spite of the potential utility of TI responses as a preventive and therapeutic intervention against a wide range of pathogens, little work has been done in this area. One reason is that, until recently, TI antibody responses have been considered exclusively short-lived and incapable of conferring long-lasting protection.

2 T Cell-Dependent and -Independent Antigens

Protein antigens such as tetanus, diphtheria, and pertussis toxins are potent immunogens. They induce generation of antigen-specific T cells that help B cells by providing cognate and noncognate interactions crucial for B cell responses. Such antigens are referred to as TD antigens. Vaccines made up of this group of antigens generate highly protective and long-lasting antibody responses (Zinkernagel 2000). Antigens that induce antibody responses without T cell help are referred to as TI antigens (Fagarasan and Honjo 2000; Vos et al. 2000). TI antigens are typically resistant to proteolysis and cannot be processed and presented to T helper cells via MHC class II, and thus are not capable of establishing cognate interaction with T cells, which can account for the restricted immunogenic nature of TI antigens (Lesinski and Westerink 2001). TI type 1 (TI-1) antigens, the prototype of which is bacterial lipopolysaccharide (LPS), activate B cells primarily by stimulating mitogenic receptors, for example Toll-like receptors (TLRs). Therefore, the antibodies generated by such stimuli are predominantly polyclonal (Vos et al. 2000). On the other hand, type 2 (TI-2) antigens such as bacterial capsular polysaccharides (PS) and other macromolecules with repetitive antigenic determinants, can induce antibody responses by primarily cross-linking the BCR of specific B cells (Vos et al. 2000; see also the chapter by J.J. Mond and J.F. Kokai-Kun, this volume).

3 B Cell Subsets

Mature B cells that generate antibody responses can be divided into four subsets: follicular (FO), marginal zone (MZ), B1a and B1b subsets (Martin and Kearney 2001; Montecino-Rodriguez and Dorshkind 2006; see also the chapter by N. Baumgarth et al., this volume). Each appears to have a distinct function in the immune system. FO B cells (also known as B2 cells), comprise the majority of B cells in the body, recirculate among the B-cell rich lymphoid follicles, participate in germinal center reactions, and mount TD antibody responses (McHeyzer-Williams 2003). MZ B cells localized to the marginal sinus of the spleen are strategically located to capture blood-borne TI particulate antigens and mount a very rapid response (Martin and Kearney 2000, 2001). The B1a subset is abundant in coelomic cavities but can also be found in the spleen (Martin and Kearney 2000, 2001). B1a cells, though developmentally distinct from MZ B cells, are also efficient in generating TI antibody responses (Berland and Wortis 2002). B1a and MZ B cell subsets recognize evolutionarily conserved antigens such as phosphorylcholine moieties present on bacteria. When mice are challenged with phosphorylcholine-expressing bacteria such as *Streptococcus pneumoniae*, these two B cell subsets unite in the development of a rapid response (Martin et al. 2001). Aspects contribute to this response include the ability of these B cell subsets differentiate into plasmablasts rapidly. Primed accessory cells of the myeloid lineage such as dendritic cells and macrophages appear to facilitate this response (Balasz et al. 2002; Martin and Kearney 2001). Nevertheless, B1a cells constitutively secrete antibodies mainly of the IgM isotype and are referred to as natural antibodies because they occur spontaneously in human cord blood, in naïve antigen-free mice, and in normal individuals in the absence of apparent antigen stimulation (Boes 2000). These antibodies have been shown to play an important role in controlling a wide range of pathogens (Boes 2000). Although the B1b subset was described years ago (Stall et al. 1992), a functional role for these cells in protective immune responses has not been found until recently (Alugupalli et al. 2004; Haas et al. 2005; Hsu et al. 2006).

4 B1b Cells

B1b cells share a number of features with B1a cells including the expression of cell-surface markers and anatomical location such as the peritoneal cavity (PerC) (Stall et al. 1992). Both subsets express high levels of membrane IgM and Mac1 (integrin α_{Mac}), low levels of IgD and B220, and lack CD23 expression. The only phenotypic marker to date that distinguishes these two B1 subsets is the expression of CD5 on B1a but not on B1b cells. Additional differences in developmental pathways (Montecino-Rodriguez et al. 2006; Tung et al. 2006) and cytokine responses also exist between these two subsets. Unlike the consistent expression of CD5 on B1a cells found elsewhere (e.g. spleen) the expression of Mac1 is restricted to the B1a and B1b cells localized to the PerC. Thus, identification of the B1b subset in sites other than PerC is limited by the lack of any other B1b-specific phenotypic marker. Interestingly, a recent analysis of PerC cells from chimeric mice revealed that an IgM^{high}, $IgD^{low/-}$, $Mac1^+$ B1 cell population emerges over time in $Rag1^{-/-}$ mice reconstituted with wild type peripheral lymph node cells (Hsu et al. 2006). The majority of their PerC B1 cell population lacked CD5 expression, indicating that lymph node cells predominantly reconstitute the B1b but not the B1a compartment. Thus, the number of B1b cells and/or B1b cell precursors in mice could be far more abundant than those found in coelomic cavities. The small number of B cells transferred compared to the numbers of B1b cells recovered in the PerC of the $Rag1^{-/-}$ chimeras likely reflects proliferation of these cells after transfer, as the B1 cell subsets are known to maintain their numbers in the PerC by their unusual self-renewing capability even after an arrest of bone marrow B lymphopoiesis (Carvalho et al. 2001; Kantor et al. 1995). Although the reasons are not completely clear, the mechanism for the preferential homing of B1a and B1b cell subsets to the PerC appears to be due to their selective attraction to CXCL13, a chemokine produced by cells in the omentum and by peritoneal macrophages (Ansel et al. 2002). Mice deficient in CCR7 and CXCR5, the receptor for CXCL13, also exhibit a PerC homing defect of B1 cells (Hopken et al. 2004; Muller et al. 2003). As a consequence of this migration impairment, the immunity of the coelomic cavity is significantly compromised in these mice (Ansel et al. 2002). Parabiosis experiments revealed that B1 cell subsets do not actively circulate compared to B2 cells (Ansel et al. 2002). Nevertheless, PerC B1 cells have the capability to leave the PerC upon appropriate stimulation. In an immunoglobulin transgenic B1 cell mouse model, it was shown that microbial products induce the migration of PerC B1 cells to mesenteric lymph nodes where they rapidly differentiate into antibody-secreting cells (Watanabe et al. 2000). Recently, it was demonstrated that Toll-like receptor (TLR)-mediated stimulation induces a rapid and transient downregulation of integrins, resulting in an efficient movement of PerC B1 cells in a chemokine-dependent fashion (Ha et al. 2006). These data and the presence of B1b cells or their precursors in the lymph node suggest that they can move in a dynamic fashion and are poised to encounter pathogens not only in the PerC, but elsewhere with high frequency and thus play an important role in protective immunity.

5 Role of B1b Cells in T Cell-Independent Responses

5.1 Immunity to Borrelia hermsii

While investigating the immune mechanism required for controlling the relapsing fever bacterium *B. hermsii*, we found for the first time a specific function for B1b cells in protective immunity. Rodents are natural reservoirs for relapsing fever bacteria, and murine infection recapitulates the critical pathophysiological aspects of the human disease (Southern and Sanford 1969; Cadavid et al. 1994; Garcia-Monco et al. 1997; Gebbia et al. 1999; Alugupalli et al. 2001a, 2001b, 2003b; see also the chapter by T.J. LaRocca and J.L. Benach, this volume). The hallmark of this infection is recurrent episodes of high-level bacteremia (~10^8 bacteria/ml blood), each caused by antigenically distinct populations of bacteria generated by DNA rearrangements of the genes encoding the variable major proteins (Barbour 1990). Remarkably, each episode is resolved rapidly within 1–3 days (Barbour and Bundoc 2001; Connolly and Benach 2001; Alugupalli et al. 2003a). TI B cell responses are necessary and sufficient for controlling *B. hermsii* infection (Barbour and Bundoc 2001; Alugupalli et al. 2003a). IgM is the most dominant isotype in TI responses (Martin and Kearney 2001). We found that mice deficient only in the secretion of IgM, but not SHM or CSR, suffer persistently high bacteremia and become moribund, indicating that IgM is the essential antibody isotype required for controlling *B. hermsii* (Alugupalli et al. 2003a). Consistent with these results, it was shown that passive transfer of IgM from convalescent mice is sufficient to confer protection (Arimitsu and Akama 1973; Yokota et al. 1997). Nevertheless, it was demonstrated that other IgG isotypes, for instance IgG3 and IgG2b, are also capable of conferring passive protection (Yokota et al. 1997). Since the control of a bacteremic episode occurs rapidly within 1–3 days, and IgM is the first isotype to be made during primary immune response to antigen/pathogen exposure (Boes 2000), we anticipated that neither the generation of other immunoglobulin isotypes by CSR nor affinity maturation of variable regions of the immunoglobulin by SHM which typically takes more than 1 week, play a critical role in this process. Activation-induced cytidine deaminase (AID) is essential for both CSR and SHM that typically occurs during a TD germinal center reaction (Muramatsu et al. 2000). Indeed, the resolution of the bacteremia and the kinetics of the *B. hermsii*-specific IgM response are indistinguishable between wild type and AID$^{-/-}$ mice, demonstrating that unmutated IgM is sufficient for controlling *B. hermsii* infection (Alugupalli et al. 2004).

To identify the B cell subsets that are capable of generating this protective IgM response we have infected IL-7$^{-/-}$ mice, which are deficient in FO B cells but not B1a, B1b, and MZ B cell subsets (Carvalho et al. 2001), and found that they control bacteremia as efficiently as wild type mice (Alugupalli et al. 2003a). Using bone marrow chimeric mice deficient in B1a cells, we have ruled out a requirement for B1a cells in the protective response against *B. hermsii* (Alugupalli et al. 2004). Severe bacterial burden in splenectomized mice during the primary bacteremic episode suggested that MZ B cells play a role in controlling *B. hermsii* (Alugupalli et al. 2003a, 2003b). Consistent with this, recently Bockenstedt and colleagues have demonstrated that MZ B cells mount anti-*B. hermsii* antibody responses (Belperron et al. 2005). Nonetheless, the rapid control of bacteremia during secondary episodes of a virulent strain or a moderate bacteremic episode by a partially attenuated strain of *B. hermsii* in splenectomized mice suggested that MZ B cells are not the only subset that contributes to protection. Having ruled out a role for the other three B cell subsets, we investigated whether B1b cells might play a role in controlling *B. hermsii* bacteremia. Mutations affecting the BCR signaling pathway result in a severe deficiency of B1 cell subsets, which are the major producers of IgM (Berland and Wortis 2002). Due to the lack of a B1b-deficient mouse model, we examined a potential role for B1b cells in X-linked immunodeficient (*xid*) mice (Khan et al. 1995; Thomas et al. 1993). Bacteremic episodes in *xid* mice are more severe than those of wild type mice, which suggests that B1b cells may play a role in controlling *B. hermsii* (Alugupalli et al. 2003a). Although *xid* mice were clearly impaired in bacterial clearance, by 4 weeks postinfection bacteremia was completely resolved, an apparent paradox if B1b cells are indeed critical for controlling bacteremia. One possible explanation is that the markedly reduced B1b subset of *xid* mice expands during this infection. In fact, Kearney and colleagues (Martin et al. 2001) have previously shown that antigen-specific TI B cell clones expand in mice

---

Fig. 1 A Expansion of the B1b cell population in *B. hermsii*-infected mice. PerC cells of uninfected or 4-week post-infected *xid* (CBA/N) mice were harvested and stained with antibodies specific for IgM, IgD, and Mac1 or CD5 and analyzed by flow cytometry (Alugupalli et al. 2003a). All B cells were first identified by IgD and IgM dual positivity (plots not shown) and were further resolved as B1 (i.e., B1a + B1b) and B1a populations by Mac1 and CD5 positivity, respectively. The percent frequency values of B1 and B1a cells, among the all PerC cells are indicated within the plots. The frequency of B1b cells was inferred from values obtained from the subtraction of the percent of B1a (CD5$^+$) cells from the percent of all B1 cells (Mac1$^+$). Data were generated by analyzing a minimum of 20,000 cells and are representative of four separate experiments.

(continued)

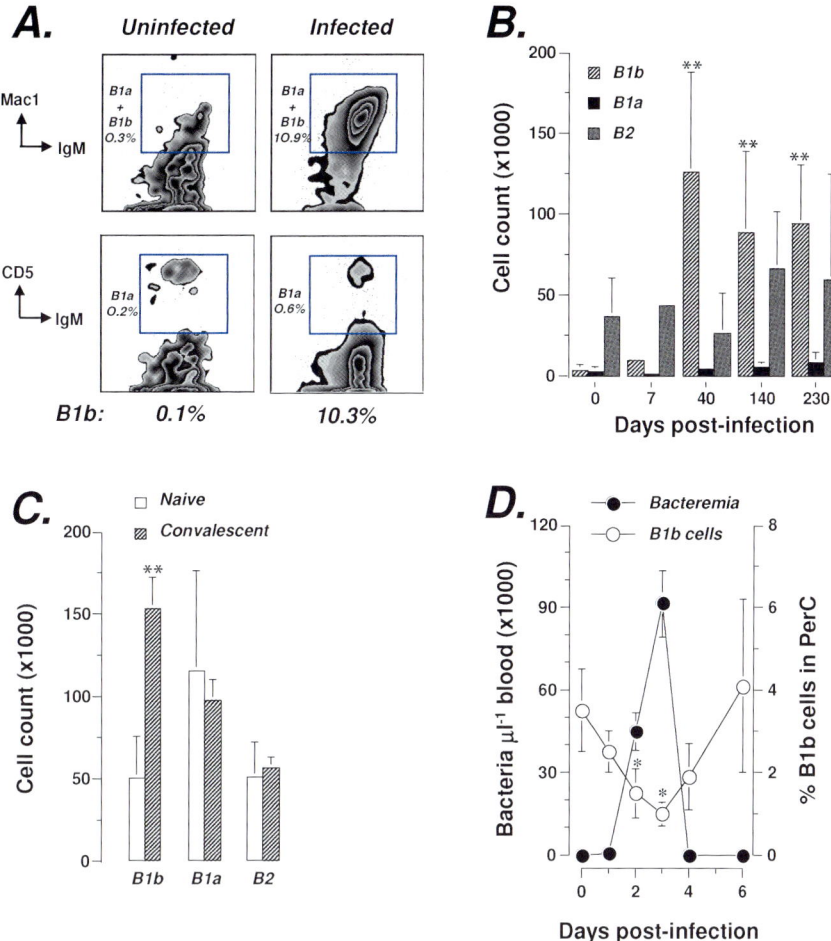

Fig. 1 (continued) The 5% contour plots are shown. **B** Persistent expansion of B1b lymphocytes in *B. hermsii*-infected mice. PerC cells of uninfected or *B. hermsii*-infected *xid* (CBA/N) mice at the indicated days postinfection (dpi) were harvested and frequencies of B1b, B1a, and B2 subsets were determined by flow cytometry. The absolute cell counts of B1b (IgM^{high}, IgD^{low}, Mac1$^+$, and CD5$^-$), B1a (IgM^{high}, IgD^{low}, Mac1$^+$, and CD5$^+$) and B2 (IgD^{high}, IgM^{low}, Mac1$^-$) were calculated as a product of their frequency and the PerC cell yield. The mean ± SD values of respective subsets of three mice at the indicated dpi are given. Significant expansion of B1b cells occurred in infected *xid* mice (**, $p<0.002$) (Alugupalli et al. 2004). **C** Expansion of B1b cells is T cell-independent. PerC cells of naïve or convalescent (40 dpi) TCR-βx$\delta^{-/-}$ mice were harvested and stained with IgM, IgD, and Mac1 or CD5 and the absolute cell numbers of B1b, B1a, and B2 were determined. The mean ± SD values of five mice are shown. Significant expansion of B1b cells occurred in infected TCR-βx$\delta^{-/-}$ mice (**, $p<0.002$) (Alugupalli et al. 2004). **D** Rapid egress of B1b cells from the peritoneal cavity (PerC) during *B. hermsii* infection. Wild type mice were infected intravenously with 5×10^4 *B. hermsii* strain DAH-p1. On each of the indicated days postinfection, three mice were killed. Bacteremia and frequency of B1b cells in the PerC were measured by microscopic counting and flow cytometry, respectively. Significant reduction of B1b cells occurred during the acute phase of the bacteremia (i.e., 2 and 3 dpi) (*, $p<0.05$)

immunized with *S. pneumoniae*. To detect a potential expansion of B1b cells, we analyzed the frequencies of PerC B cells in *xid* mice. Remarkably, the frequency of B1b cells as defined by the surface markers IgM^{high}, IgD^{low}, Mac1$^+$ and CD5$^-$ in convalescent *xid* mice is significantly greater than in uninfected *xid* mice (Fig. 1A) and comparable to the frequency of B1 cells in wild type mice (Alugupalli et al. 2003a). This expansion was selective for B1b cells but not for B1a or B2 cell subsets (Fig. 1B) and occurred regardless of whether the *xid* mice were infected with a highly virulent or a partially attenuated strain of *B. hermsii*. These findings indicate a role for B1b cells in the control of both high and moderate bacteremic episodes (Alugupalli et al. 2003a). Since IgM is essential for controlling this infection and B1b expansion was detected as early as by 7 days postinfection (Fig. 1B), we predicted that *xid* mice would be capable of mounting an anti-*B. hermsii* IgM response, despite their inherent deficiency in preimmune IgM levels. As predicted, with some delay, *xid* mice generated a specific IgM response coincident with the resolution of bacteremia (Alugupalli et al. 2007).

A significant increase in B1b cell numbers also occurs in TCR-$\beta\chi\delta^{-/-}$ mice, demonstrating that this expansion is a TI process (Fig. 1C). Remarkably, the convalescent TCR-$\beta\chi\delta^{-/-}$ were as resistant as that of the wild type mice to reinfection, suggesting that a memory-like response can be generated by B1b cells independent of a germinal center reaction (Alugupalli et al. 2004). Indeed, we eliminated a role for two very important TD germinal center events, namely SHM and CSR in controlling *B. hermsii* infection (Alugupalli et al. 2004). Movement of PerC B1 cells is crucial for their protective immune responses (Ansel et al. 2002; Ha et al. 2006; Hopken et al. 2004; Muller et al. 2003). We therefore anticipated that B1b cells leave the PerC to differentiate into IgM secreting plasma cells that are critical for the *B. hermsii* immunity. Upon infection with *B. hermsii*, a rapid egress of B1b cells from the wild type mouse PerC was also observed. The magnitude of the egression was directly proportional to the bacterial burden (Fig. 1D). Interestingly, soon after resolution of bacteremia, B1b cell numbers in the PerC were rapidly restored to basal levels, suggesting an active turnover and that these cells are capable of participating in protective immunity in the blood stream (Fig. 1D).

To directly determine whether B1b cells are capable of providing long-lasting immunity, we transferred B1b cells from convalescent mice into Rag1$^{-/-}$ mice that are otherwise completely incapable of eliminating *B. hermsii* (Alugupalli et al. 2003a). Rag1$^{-/-}$ mice reconstituted with B1b cells controlled *B. hermsii* bacteremia, demonstrating for the first time a crucial role for B1b cells in protective immunity (Alugupalli et al. 2004). The fact that naïve mice suffer recurrent bacteremic episodes indicates that naïve B1b cells are functionally less effective than convalescent B1b cells. As expected, transfer of the same number of naïve B1b cells to Rag1$^{-/-}$ mice conferred only partial protection (Alugupalli et al. 2004). Consistent with the requirement for IgM in the clearance of bacteremia, the transferred B1b cells in Rag1$^{-/-}$ mice confer protection by mounting a specific IgM response against *B. hermsii*, and the generation of this IgM is coincident with the resolution of bacteremia. B1b cells do not mount an antibody response in the absence of specific stimulation, indicating that they maintain a quiescent state like conventional memory B cells (Alugupalli et al. 2004). This property of B1b cells is not unique to the *B. hermsii* infection, and recently has been extended to another important human pathogen, *S. pneumoniae*.

5.2 Immunity to Streptococcus pneumoniae

A number of clinically important pathogens, including *S. pneumoniae*, *Haemophilus influenzae*, and *Neisseria meningitidis* express PS capsules (Lesinski and Westerink 2001). Some pathogens associated with biological threats such as *Bacillus anthracis*, the etiological agent of anthrax, express capsules made of poly-γ-d-glutamic acid (Wang and Lucas 2004). Due to the restricted immunogenicity for TD responses, the masking of potentially immunogenic TD protein antigens, and the provision of serum resistance, capsules enable the persistence of these pathogens in the host and eventually cause serious diseases ranging from bacteremia to meningitis (Kelly et al. 2004; Scorpio et al. 2007). A critical factor in preventing these infections from becoming systemic is the ability of B cells to rapidly generate protective antibodies against capsules. As mentioned earlier, bacterial capsules are TI-2 antigens and antibodies to capsules are highly protective. In fact, the introduction of conjugate vaccine against the capsular PS (Hib) had a major impact in reducing the incidence of *H. influenzae* type b disease worldwide (Kelly et al. 2004).

Using a well-established murine model of *S. pneumoniae* infection in mice sufficient or deficient in CD19, Haas and colleagues found that CD19, though an important molecule for B cell development is, surprisingly, not necessary for generating protective IgM or IgG3 responses against the capsular PS of *S. pneumoniae* (Haas et al. 2005). $CD19^{-/-}$ mice have a severe deficiency in B1a cells and MZ B cells but not B1b cells, suggesting that this subset is responsible for the observed anti-PS specific antibody response. Conversely, CD19 transgenic (CD19Tg) mice that are severely deficient in B1b cells but not B1a or MZ B cells were incapable of mounting an antibody response to capsular PS (Haas et al. 2005). Immunization of $CD19^{-/-}$ but not CD19Tg mice with capsular PS protects mice from a lethal *S. pneumoniae* challenge. These correlations suggest that B1b cells might be involved in generating long-lasting anti-PS responses. To identify the B cell subsets responsible for the anti-PS antibodies, Haas and colleagues transferred wild type B1b, B1a, or MZ B cells into $Rag1^{-/-}$ mice. None of the reconstituted mice spontaneously generated anti-PS antibodies over a period of 4 weeks. However, upon PS immunization only B1b-reconstituted mice mounted an exuberant anti-PS IgM and IgG3 response (Haas et al. 2005). Unreconstituted mice died as early as 2 days postinfection when challenged with as few as 100 colony-forming units of *S. pnuemoniae*. In striking contrast, greater than 75% of $Rag1^{-/-}$ mice reconstituted with B1b cells survived this lethal dose (Haas et al. 2005). These data demonstrate yet another example of B1b cell-mediated immunity to a clinically important bacterial pathogen. In the pneumococcal infection mouse model B1a cell subsets were also shown to play an important complementary role in protective immunity by virtue of their natural antibody repertoire. Unlike constitutively secreted natural antibodies, antibodies generated by B1b cells are inducible only by specific antigen stimulation (Haas et al. 2005), as in the case of *B. hermsii*-specific B1b cell response (Alugupalli et al. 2004). To date, two independent studies, using clinically distinct murine infection systems, reveal a long-lasting and antigen-specific B1b cell response (Alugupalli et al. 2004; Haas et al. 2005). In both cases, the function of B1b cells was revealed unpresumptuously. Therefore, it is tempting to consider an important role for B1b cells in protection against other infections.

5.3 Response to NP-Ficoll, a Model TI-2 Antigen

Due to the division of labor within TI B cell subset responses during an infection and the host or tissue tropism of bacterial pathogens, it is not easy to dissect the role of B1b cells in other infection systems. Moreover, the lack of a B1b cell-specific marker currently poses a challenge in tracking B1b cell responses in compartments such as the spleen, lymph nodes, and bone marrow. Nevertheless, while exploring the identity of B cell populations involved in the response to 4-hydroxy-3-nitrophenyl acetyl (NP), a frequently used hapten for studying antigen-specific antibody responses in mice, MacLennan and colleagues revealed an involvement for B1b cells in anti-NP responses (Hsu et al. 2006). NP-conjugated polysaccharide Ficoll is a widely used model antigen for studying TI-2 responses. It is known that extra-follicular antibody responses induced by immunization of NP-Ficoll persist for long periods of time (de Vinuesa et al. 2000). Adoptive transfer experiments were performed using Rag1$^{-/-}$ mice reconstituted with wild type mouse PerC cells which were depleted of either B2 or B1a cells. These mice generated normal levels of NP-specific IgM and IgG3 antibodies (Hsu et al. 2006). In contrast, Rag1$^{-/-}$ mice reconstituted with B1a and B1b cell-depleted PerC cells failed to generate NP-specific antibody responses, indicating B1b cells are involved in this response (Hsu et al. 2006). In support of the specific induction of B1b cell-derived IgM seen in *B. hermsii* and *S. pneumoniae* infection models (Alugupalli et al. 2004; Haas et al. 2005), NP-specific B1b cells did not generate an anti-hapten response spontaneously (Hsu et al. 2006). The magnitude of the anti-NP response increased dramatically following NP-Ficoll immunization (Hsu et al. 2006), revealing once again a requirement for specific-antigen stimulation in the B1b cell response (Alugupalli et al. 2004; Haas et al. 2005; Hsu et al. 2006). Surprisingly, Rag1$^{-/-}$ mice reconstituted with lymph node B cells also generated NP-specific IgM and IgG3 responses as efficiently as the PerC cell-reconstituted Rag1$^{-/-}$ mice (Hsu et al. 2006). Analysis of PerC cells isolated from lymph node cell-reconstituted Rag1$^{-/-}$ mice revealed a robust generation of B1b cells but not B1a cells. Approximately 2% of lymph node B220$^+$ cells are IgM^{high} and IgD^{low}, which are characteristics of B1b cells. These results indicate that B1b cells and/or B1b cell precursors are present in lymphoid compartments and are poised to contribute to TI-2 antigen-specific responses. Recently, it was demonstrated that IgM memory B cells are generated upon

NP-Ficoll immunization and such cells can be detected in spleen (Obukhanych and Nussenzweig 2006). The phenotype of these B cells is distinct from that of MZ B and FO B cells. It is possible that these cells might represent splenic B1b cells (Obukhanych and Nussenzweig 2006).

6 Activation of Antigen-Specific T Cell-Independent B Cell Responses

Antigens driving B1b cell responses appear to be heterogenous. For example, one *B. hermsii* antigen targeted by B1b cells is a protein implicated in virulence of this pathogen (M.J. Colombo and K.R. Alugupalli, unpublished data). In the case of *S. pneumoniae*, it is a carbohydrate (Haas et al. 2005), whose expression is implicated in a variety of immune evasion strategies. Thus, the recognition of biochemically different bacterial products and even synthetic haptens such as NP (Hsu et al. 2006) by B1b cells clearly indicates that the B1b repertoire is capable of responding to a wide spectrum of antigens. Unbiased analysis of the CDR3 regions of VH genes revealed that unlike B1a cells, B1b cells have high junctional diversity that is comparable to B2 cells (Kantor et al. 1997; Tornberg and Holmberg 1995). Furthermore, immunization with NP-Ficoll can result in low levels of SHM in the VH regions even without T cell help (Toellner et al. 2002). This process presumably occurs by BCR cross-linking induced AID expression without the requirement for CD40-CD40L (Faili et al. 2002; Weller et al. 2001). This could also explain the existence of other antibody isotypes such as IgG3 by CSR during TI responses (Martin and Kearney 2001). However, unlike conventional B2 cells, the affinity and overall Ig repertoire of B1b cell-derived antibodies may be limited due to the lack of an involvement with TD germinal center reactions.

6.1 Role of Btk in T Cell-Independent B Cell Responses

Although antigen-specific B1b cells respond to various TI antigens, for example pneumococcal PS, hapten NP, and *B. hermsii* surface protein (Haas et al. 2005; Hsu et al. 2006) (M.J. Colombo and K.R. Alugupalli, unpublished data), the mechanism of activation of this response can vary. *Xid* mice, which carry a point mutation in Bruton's tyrosine kinase (Btk), a cytoplasmic kinase belonging to the Tec family of kinases is crucial for BCR-mediated activation. B cells of these mice cannot respond to anti-BCR cross-linking and are therefore impaired primarily in response to TI-2 antigens such as NP-Ficoll (Amsbaugh et al. 1972; Khan et al. 1995; Scher et al. 1975; Thomas et al. 1993; see also the chapter by J.J. Mond and J.F. Kokai-Kun, this volume). In fact, these mice cannot generate protective antibody responses to capsules of *S. pneumoniae* or *B. anthracis* (Briles et al. 1981; Wang and Lucas 2004). Mice that have a targeted deletion in the *Btk* gene recapitulate the *xid* mouse phenotype

(Khan et al. 1995). Surprisingly, the phenotypes seen in these mice are not identical to X-linked agammaglobulinemia (XLA) patients, the human equivalent of *xid*. Mice deficient in Tec, a kinase belonging to the same family as Btk, have no major phenotypic alterations of the immune system (Ellmeier et al. 2000). Interestingly, mice deficient in both Btk and Tec exhibit a phenotype that is similar to human XLA (Ellmeier et al. 2000). Although the exact reasons for these differences is not clear, it was demonstrated that retroviral-mediated transfer of the human *Btk* gene corrects a number of functions, including TI-2 responses in Btk and Tec double knockout mice, indicating a functionally conserved and crucial role of Btk in both mice and humans (Yu et al. 2004). *Xid* mice respond to LPS, a model TI-1 antigen (Hiernaux et al. 1983; Mosier et al. 1977). The moderate impairment of *xid* mouse B cells to LPS stimulation (Khan et al. 1995) is likely due to a potential role for Btk in TLR-mediated signaling rather than its role in BCR-mediated signal (Jefferies et al. 2003; Jefferies and O'Neill 2004), since the IgM response induced by LPS is polyclonal and nonspecific (Andersson et al. 1972; Mosier et al. 1976). In support of this, it was shown that mice deficient in MyD88, involved in TLR- but not BCR-mediated signaling, exhibit impaired LPS-induced B cell activation (Kawai et al. 1999). In fact, these mutant mice generate normal anti-NP-Ficoll antibody responses (Schnare et al. 2001). These data together reveal the critical role for Btk in TI-2 responses.

Although *xid* mice are deficient in B1 cells, as previously mentioned (see Sect. 5.1), the impairment of a specific antibody response to TI-2 antigens is not entirely due to a deficiency of these B cell subsets. Increasing the B cell numbers in *xid* mice by blocking apoptosis of B cells with the transgenic expression of Bcl2, does not restore B cell functions, including TI-2 responses in *xid* mice (Woodland et al. 1996). Transgenic expression of IL-9 or injection of this cytokine induces selective expansion of B1b cells in wild type mice (Vink et al. 1999). Such a transgene can even correct the B1b cell deficit of *xid* mice but does not restore the specific antibody response to pneumococcal capsular PS or NP-Ficoll (Knoops et al. 2004). In contrast, a Btk transgene restores NP-Ficoll and antiviral TI-2 responses of *xid* mice in a dose-dependent fashion (Pinschewer et al. 1999; Satterthwaite et al. 1997). Furthermore, the defective anti-PS response of *xid* mice is not attributed to lack of a specific VH gene family involved in TI-2 responses (Feng and Stein 1991; Selinka and Bosing-Schneider 1988). These data suggest that the impaired TI response in *xid* mice is due to a compromised BCR-mediated signal rather than a deficiency in B1b cells or a potential deficiency in the BCR repertoire itself.

6.2 Role of Co-stimulatory Signals in T Cell-Independent B Cell Responses

6.2.1 CD40 and Its Ligand CD40L

Despite a BCR-mediated signaling defect, *xid* or Btk$^{-/-}$ mice generate near normal responses to TD antigens (Khan et al. 1995). This response is likely mediated by CD40-CD40L co-stimulation provided by T cells, since mice deficient in both Btk and CD40 are severely compromised for TD antibody responses (Khan et al. 1997). In fact, in vitro co-stimulation of *xid* B cells with CD40 obviates the need for Btk in BCR-mediated signal as detected by cell cycle progression and nuclear translocation of NF-κB, a transcription factor involved in a variety of B cell responses (Mizuno and Rothstein 2003; Mizuno and Rothstein 2005). Consistent with this, injection of anti-CD40 antibodies as a surrogate for CD40L restores immune responses to NP-Ficoll in *xid* mice (Dullforce et al. 1998; Vinuesa et al. 2001). In striking contrast to that of typical TI-2 responses induced by capsular PS or NP-Ficoll, the antibody response to *B. hermsii* is clearly different. For example, unlike pure TI-2 antigens, an active infection is expected to engage several immune activation pathways. Although *xid* mice suffer more severe bacteremia than wild type mice, they control all episodes of *B. hermsii* bacteremia and the resolution of the infection coincides with an expansion of B1b cells in these mice (see Sect. 5.1) (Alugupalli et al. 2003a). *Xid* mouse B cells are defective in homeostatic proliferation (Cabatingan et al. 2002; Woodland and Schmidt 2005), hence this B1b cell expansion is likely to be antigen-driven. In support of this, *xid* or $Btk^{-/-}$ mice generate an antigen-specific IgM response capable of controlling infection (Alugupalli et al. 2007). The magnitude of the specific IgM response in wild type mice is comparable to that of TCR-βx$\delta^{-/-}$ mice, implying an unlikely role for CD40-CD40L-mediated co-stimulation, provided by T cells in this infection (Alugupalli et al. 2007) (Fig. 2A). The observation that $CD40L^{-/-}$ mice generate a robust TI IgM response to the related bacterium *B. burgdorferi* supports this notion (Fikrig et al. 1996). Although the physiological relevance of CD40-CD40L interaction in TI responses is not clear, the above data indicate that a role exists for alternate pathways of activation during TI responses.

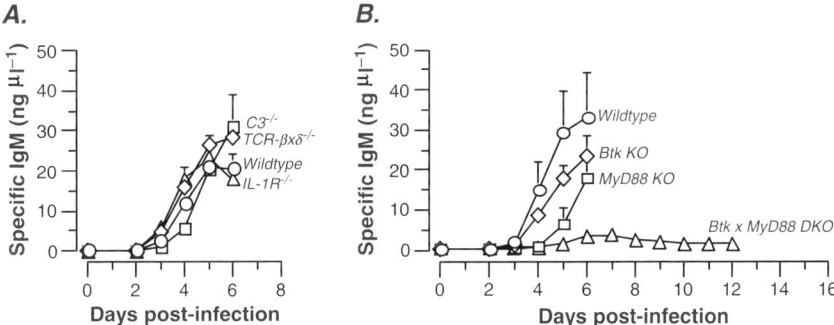

Fig. 2 Kinetics of the *B. hermsii*-specific IgM response in various knockout mice. **A** Wild type (n=3), T cell-deficient (TCR-βx$\delta^{-/-}$; n=3), $C3^{-/-}$ (n=5), or IL=1$R^{-/-}$ (n=5) mice; **B** Wild type (n=3) or mice deficient in either Btk (Btk KO; n=3) or MyD88 (MyD88 KO; n=3) or both (Btk x MyD88 KO; n=4) were infected intravenously with 5×10^4 *B. hermsii* strain DAH-p1. *B. hermsii*-specific IgM responses in the blood on the indicated days postinfection were measured by ELISA and mean ± SD values are shown (Alugupalli et al. 2007)

6.2.2 Complement Component C3 and Its Receptor CR1/2

The high expression of CR1/2, the C3 receptor on B cells (Zandvoort and Timens 2002), is necessary for specific targeting of C3 conjugated TI-2 antigens such as pneumococcal capsular PS to MZ B cell (Breukels et al. 2005; Guinamard et al. 2000). Mice deficient in C3 have reduced responses to TI-2 antigens (Guinamard et al. 2000). Some viral pathogens also behave like TI antigens and induce TI-2 like responses in vivo (Fehr et al. 1998; Ochsenbein et al. 1999; Szomolanyi-Tsuda et al. 2001; Szomolanyi-Tsuda and Welsh 1998). Studies of vesicular stomatitis virus have demonstrated a correlation between antigen repetitiveness (TI-2 like arrangement) and the degree to which B-cell activation is independent of T cells. The rigidly structured Poliovirus efficiently induces neutralizing IgM antibodies independent of T cell help (Fehr et al. 1998). For example, Poliovirus and Vaccinia virus expressing vesicular stomatitis virus glycoproteins do not induce antibody responses in *xid* mice (Pinschewer et al. 1999; Szomolanyi-Tsuda et al. 2001). However, as mentioned earlier, a Btk transgene compliments these responses in *xid* mice, suggesting that BCR signaling is the limiting component (Pinschewer et al. 1999). However, impaired TI responses to these viral pathogens, in $C3^{-/-}$ or $CR2^{-/-}$ mice, implied that simultaneous engagement of C3-coupled antigen to CR1/2 and BCR promotes B cell stimulation (Ochsenbein et al. 1999; Szomolanyi-Tsuda et al. 2006). In fact, priming of human B cells with C3 engagement to CR1/2 enhances anti-IgM-mediated B cell activation even to very low levels of BCR cross-linking (Carter and Fearon 1989; Carter et al. 1988). This co-stimulatory pathway seems to be more appropriate for the activation of MZ B cells than B1b cells. For instance, B1b but not MZ B cells are crucial for the clearance of *B. hermsii* and $C3^{-/-}$ mice control this infection efficiently (Connolly and Benach 2001; Connolly et al. 2004). The kinetics of the anti-*B. hermsii* IgM response in $C3^{-/-}$ mice is comparable to that of wild type mice (Fig. 2A). This result is in agreement with the C3-independent IgM-mediated killing of relapsing fever bacteria (Connolly and Benach 2001; Connolly et al. 2004).

Xid B cells respond to TI-2 antigens when co-stimulated by IL-1 (Couderc et al. 1987). However, we found that $IL-1R^{-/-}$ mice generate a robust IgM response to *B. hermsii* (Fig. 2A) (Alugupalli et al. 2007). Our results on the pathogen-specific IgM response in *xid* or $Btk^{-/-}$ mouse infections suggested that even in the absence of normal BCR-mediated signaling, *B. hermsii* is capable of rapidly activating antigen-specific B cells, presumably by stimulating other signaling pathways such as the TLR-mediated pathway. In fact, when *xid* B cells are co-stimulated with TLR ligands, they respond to BCR cross-linking (Couderc et al. 1987).

6.2.3 Toll-Like Receptors

TLRs play important roles in activation of the immune system (Takeda and Akira 2005). Importantly, TLR9 has been shown to activate auto-reactive B cells

(Leadbetter et al. 2002) and memory B cells (Bernasconi et al. 2002), indicating a possible role for TLR9 in the development of rapid and long-lasting TI-2 antibody responses. Since *B. hermsii* is a prokaryote, it is expected to contain a high frequency of CpG DNA, the ligand for TLR9. Despite these expectations, TLR9$^{-/-}$ mice generate rapid anti-*B. hermsii* IgM responses (Alugupalli et al. 2007). *B. hermsii* possess other potential TLR ligands such as lipoproteins, the ligands for TLR2 (Shang et al. 1998) that could redundantly activate distinct members of the TLR family, thereby restoring responses in *xid* mice. Indeed, mice deficient in MyD88, a common cytoplasmic adaptor for all TLRs except TLR3, exhibited a significantly delayed anti-*B. hermsii* IgM response (Fig. 2B). Analysis of individual TLR knockout mice revealed important roles for TLR1 and TLR2 in anti-*B. hermsii* IgM responses (Alugupalli et al. 2007). CD14 is known to augment TLR2-mediated responses and as expected, CD14$^{-/-}$ mice have impaired antibody responses to *B. hermsii* (Alugupalli et al. 2007). CD14 is not only involved in TLR2 signaling but also in TLR4-mediated signaling. Although *B. hermsii* is not a Gram-negative bacterium and hence does not contain LPS, other evolutionarily distant microbial components such as glycolipids of Treponema (Schroder et al. 2000), pneumolysin of *S. pneumoniae* (Malley et al. 2003), and fusion protein of respiratory syncytial virus (Kurt-Jones et al. 2000) are recognized by TLR4. This may explain the delayed and reduced *B. hermsii*-specific IgM response in TLR2$^{-/-}$ mice. The magnitude of the specific IgM response in TLR4$^{-/-}$ mice was also somewhat lower than wild type mice (Alugupalli et al. 2007). As expected, this potential TLR4-mediated response did not resemble an LPS-induced response, since mice deficient in MD2, a protein crucial for the LPS response (Nagai et al. 2002), generated normal antibody responses (Alugupalli et al. 2007). These results indicate that *B. hermsii* may also signal through TLR4 in addition to TLR1 and TLR2, by a mechanism distinct from that of LPS-induced stimulation. Mice deficient in TLR3, which is involved in recognizing double-stranded RNA, a signature for viral rather than bacterial infection, generated IgM responses comparable to wild type mice (Alugupalli et al. 2007).

Due to the redundancy in TLR function, we tested whether MyD88 provides synergistic or partially redundant functions in triggering an antibody response to *B. hermsii* in *xid* mice. Indeed, mice deficient in both Btk and MyD88 are severely compromised in IgM production and bacterial clearance but not TD antibody responses that utilize CD40-CD40L (Alugupalli et al. 2007). Despite a normal response both in terms of kinetics and magnitude of specific IgM or IgG to the model TD antigen NP-conjugated chicken gammaglobulin, Btk x MyD88$^{-/-}$ mice are unable clear *B. hermsii* even transiently, highlighting the critical role for TI responses in protective immunity (Alugupalli et al. 2007) (Fig. 2B). Although B1b cells are capable of efficiently recognizing chemically distinct antigens, these data suggest the involvement of a co-stimulatory mechanism in B1b cell activation during TI responses. Understanding the sources of TLR signaling and how they regulate B1b cell function may help develop strategies to restore efficient responses in individuals with impaired TI responses.

7 Impaired T Cell-Independent B Cell Responses

Children and the elderly respond poorly to TI antigens, and as a consequence suffer severe and recurrent infections by both encapsulated and nonencapsulated bacteria (Kelly et al. 2004, 2005; Lesinski and Westerink 2001; see also the chapter by JJ. Mond and J.F. Kokai-Kun, this volume). A number of possibilities could account for this impairment, some of which are reminiscent of the defects seen in *xid* mice, suggesting that the *xid* mouse is an appropriate model for studying impaired TI responses in children. Similar to the BCR-mediated activation defect of *xid* B cells (Couderc et al. 1987), it has been shown that the B cells of neonatal, very young, and aged wild type mice exhibit an activation defect to PS antigens and multivalent membrane Ig crosslinking, a mimic of TI-2 antigen-induced stimuli (Chelvarajan et al. 1998, 1999; Snapper et al. 1997). This impairment of the surface IgM cross-linking and NP-Ficoll responses in neonatal B cells can be corrected by a number of the above-mentioned (see Sect. 6) co-stimulatory signals for *xid* B cells (Couderc et al. 1987; Vinuesa et al. 2001) such as CD40-L, TLR ligands, IL-1, and IL-6 (Chelvarajan et al. 1998, 1999; Snapper et al. 1997). A defect in accessory cell-mediated co-stimulation for neonatal and aged mouse B cell activation has also been proposed to explain the poor responses of children and the elderly. These accessory cells are mainly of the myeloid lineage, such as monocytes and macrophages, but not T cells (Bondada et al. 2000; Chelvarajan et al. 2004, 2005, 2006; Landers et al. 2005; Yan et al. 2004a, 2004b). In fact, compared to adult human monocytes, neonatal monocytes express significantly lower levels of MyD88 protein, which is required for IL-1 and almost all TLR-mediated signaling (Yan et al. 2004b). In agreement with these signaling defects, we have recently demonstrated that mice deficient in both MyD88 and Btk are severely compromised for TI antibody responses (see Sect. 6.2.3).

The lack of appropriate B cell subsets and/or IgM memory B cells could also explain the impaired TI responses in the young (Kruetzmann et al. 2003; Zandvoort et al. 2001). For example, the lack of MZ B cell development and the low expression of CR1/2 on MZ B cells in the spleens of children under 2 years of age has been suggested as a reason for their impaired responses to PS (Zandvoort and Timens 2002). Nevertheless, upon immunization with NP-Ficoll, pyk-$2^{-/-}$ mice, deficient in MZ B cells but not B1 cells, generate IgM and IgG3 antibody responses more than two orders of magnitude higher than preimmune levels (Guinamard et al. 2000). This implies that other B cells, likely B1b cells, play an important role in this TI-2 response. In agreement with this possibility, two independent studies reviewed here (see Sects. 5.2 and 5.3) provided direct evidence using Rag$1^{-/-}$ chimeras that B1b cells generate the majority of the anti-TI-2 IgM and IgG3 responses to NP-Ficoll and PS (Haas et al. 2005; Hsu et al. 2006). It is interesting that the splenic localization of PS conjugates can also be independent of the presence of C3 (Breukels et al. 2005), which could explain the significant anti-NP-Ficoll response in pyk-$2^{-/-}$ mice (Guinamard et al. 2000). In agreement with this, mice deficient in CD19, the signal-transducing moiety of the C3 receptor CR1/2, generate

B1b cells normally and mount efficient protective antibody responses to capsular PS (Haas et al. 2005). In fact, C3$^{-/-}$ mice also generate normal IgM responses to *B. hermsii* (Fig. 2A) (Connolly and Benach 2001; Connolly et al. 2004). While it is not clear that all distinctions between B cell subsets defined in the mouse hold true for humans, these data also suggest that children lack sufficient protection from functional IgM-secreting (memory) B1b cells. Such a lack in the rapid as well as long-lasting responses to TI antigens by B1b cells may explain recurrent infections by encapsulated bacterial pathogens in children.

8 Memory B1b Cells

The functional definition of immunological memory is the ability of the immune system to respond more rapidly and effectively to pathogens that have been encountered previously and reflect the preexistence of clonally expanded populations of antigen-specific lymphocytes (Janeway et al. 2004). The long-term immunity provided by B1b cells functionally resembles that of canonical B cell memory; however, it is generated and maintained in the complete absence of T cells (Alugupalli et al. 2004).

The B1b cell responses to three distinct TI antigens reviewed here (Sects. 5.1, 5.2, and 5.3) have revealed strikingly common characteristics to memory B cell responses. Such characteristics include (1) the generation of B1b cell responses that do not require continuous B lymphopoiesis (Alugupalli et al. 2003a; Carvalho et al. 2001); (2) B1b cells maintain their numbers by self-renewal (Kantor et al. 1995); (3) B1b cells maintain a quiescent state in terms of their differentiation into plasma cells, and their activation requires specific-antigenic stimulation (Alugupalli et al. 2004; Haas et al. 2005; Hsu et al. 2006); and (4) B1b cells generate long-lasting antibody responses (Alugupalli et al. 2004; Haas et al. 2005; Hsu et al. 2006). In the *B. hermsii* infection system, expanded B1b cells persist for a remarkably long time in convalescent mice (Fig. 1B) and upon challenge, expanded B1b cells rapidly differentiate into antibody secreting cells (Alugupalli et al. 2004). While naïve B1b cells also generate specific IgM, the magnitude of the response is significantly lower compared to that of expanded B1b cells. The kinetics of the IgM response is also considerably delayed (K.R. Alugupalli, unpublished data). These properties indicate that a subset of B1b cells behave like memory B cells. Reasons that account for the evolution of the enhanced protection or antibody responses by such a subset may include an expansion of antigen-specific B1b cells. In addition, antigenic stimulation may have conferred upon immune or convalescent mouse B1b cells a distinct property that naïve B1b cells do not possess, similar to those that distinguish TD memory B cells from naïve B cells, as illustrated in Fig. 3.

The robust binding of B1b cell-derived IgM to *B. hermsii* without SHM (Alugupalli et al. 2004) suggests an increased quantity of antigen-specific B1b cells in convalescent mice rather than an affinity maturation of the VH regions of specific B1b clones. Interestingly, the VH sequences of the majority of the Ig

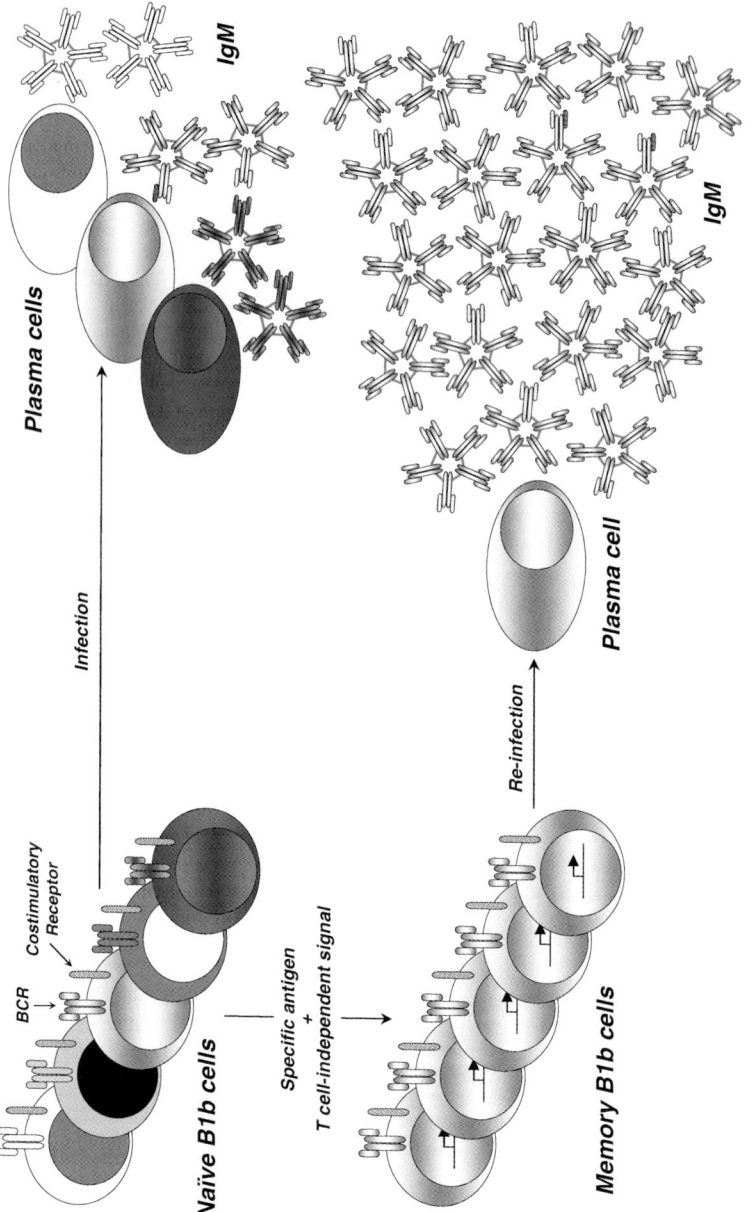

Fig. 3 A model for the efficient protection conferred by B1b cells to pathogens in convalescent mice. Upon infection or exposure to a specific T cell-independent antigen, B1b cells of varying affinities generate a short-lived IgM response. In this primary response, a division of labor by appropriate MZ B and/or B1a subsets (not shown in the figure) may also contribute to the IgM response. While a significant fraction of this IgM may recognize the pathogen, only some of this response is protective during primary infection. On the other hand, a subset of B1b cells driven

repertoire of NP-Ficoll immunized mice have no SHM, and these antibodies are of the IgM isotype (Maizels and Bothwell 1985; Maizels et al. 1988). This unmutated IgM response is not unique to the carrier, Ficoll. In fact, mice deficient in BCL6, which cannot develop a germinal center reaction, are capable of mounting hapten-specific antibody responses when immunized with NP-conjugated to chicken gammaglobulins as a carrier, a frequently used model TD antigen. The VH regions of the NP-specific IgM memory B cells in BCL6$^{-/-}$ mice are also unmutated. These results indicate that generation of unmutated IgM memory occurs without germinal center formation and is specific to the antigen driving such a response (Toyama et al. 2002). In fact, such an unmutated IgM response is sufficient for conferring immunity even to a rapidly replicating bacterial pathogen, since TI control of *B. hermsii* occurs in AID$^{-/-}$ mice. Furthermore, we have also shown that convalescent AID$^{-/-}$ mice are also resistant to reinfection, demonstrating that the unmutated IgM memory of B1b cells is functional (Alugupalli et al. 2004).

9 Concluding Remarks

To date three distinct model systems have revealed that the B1b cell subset mounts the majority of the antigen-specific TI response (Alugupalli et al. 2003a, 2004; Haas et al. 2005; Hsu et al. 2006). This B1b cell subset is also capable of generating a unique response that is functionally but not developmentally similar to the classical memory B cell response. The generation of rapid and long-lasting IgM induction from the B1b cell subset may not be an unnecessary complexity in the seemingly linear conventional TD B cell memory development, as suggested recently (Tarlinton 2006). The existence of B1b cell memory does not appear to be a redundant arm of the memory B cell compartment, as B1b cell memory has already been shown to play an indispensable role in two clinically important bacterial infections (Alugupalli et al. 2004; Haas et al. 2005). In fact, TD responses do not contribute to protective immunity in these systems. The dichotomy in the development of B cell memory, such as the generation of TI memory B1b cells and canonical TD memory B cells seem to imply the humoral immune system's inherent strategy to cover two major categories of immunogens, the TI and TD antigens, respectively. The division of labor by these two distinct arms of B cell memory

Fig. 3 (continued) by specific antigens expand with the help of a T cell-independent co-stimulatory signal (see Sect. 6), persist in convalescent mice with higher frequency, and maintain a quiescent state in terms of their further differentiation into plasma cells. Such B1b cells acquire a functional memory B1b cell phenotype. Compared to their naïve counterparts, the ability of memory B1b cells to mount a heightened IgM response, upon antigen exposure, results in efficient protection from reinfection. Factors that may contribute to this rapid response include an altered transcriptomic profile and surface expression of other signaling receptors. Since AID can be induced by BCR cross-linking and TLR stimulation, IgM secreting B1b cells could also switch to other isotypes, such as IgG3

could ensure achieving sterilizing immunity to a wider range of pathogens in an overlapping and timely fashion. The rapid and antigen-specific IgM response enhances the germinal center reaction by forming antigen–IgM complexes and facilitates an accelerated development of high-affinity antibody response to TD antigens, suggesting a role for B1b-derived IgM in the development of canonical B cell memory (Corley et al. 2005). In addition, B1b cell-derived TI responses may also be beneficial in decreasing the incidence of morbidity associated with opportunistic pathogens in individuals with decreased T cell counts and T cell dysfunction (Chinen and Shearer 2002). For instance, a selective reduction in CD27[+] conventional memory B cells in drug naïve and in highly active anti-retroviral therapy-treated HIV-1-infected individuals compromises their TD responses and increases susceptibility to a number of opportunistic infections (Chong et al. 2004). Elucidation of the developmental pathways involved in B1b cell generation, activation, expansion, and the long-term maintenance of antigen-specific B1b cells may provide new approaches to the induction of long-lasting protective antibody responses in individuals impaired in TI and/or TD responses.

Acknowledgements I thank the support from NIH/NIAID RO1 AI065750. I dedicate this review to the memory of my brother, Dr. Srinivas Alugupalli.

References

Alugupalli KR, Michelson AD, Barnard MR, Leong JM (2001a) Serial determinations of platelet counts in mice by flow cytometry. Thromb Haemost 86:668–671

Alugupalli KR, Michelson AD, Barnard MR, Robbins D, Coburn J, Baker EK, Ginsberg MH, Schwan TG, Leong JM (2001b) Platelet activation by a relapsing fever spirochete results in enhanced bacterium-platelet interaction via integrin $\alpha IIb\beta 3$ activation. Mol Microbiol 39:330–340

Alugupalli KR, Gerstein RM, Chen J, Szomolanyi-Tsuda E, Woodland RT, Leong JM (2003a) The resolution of relapsing fever Borreliosis requires IgM and is concurrent with expansion of B1b lymphocytes. J Immunol 170:3819–3827

Alugupalli KR, Michelson AD, Joris I, Schwan TG, Hodivala-Dilke K, Hynes RO, Leong JM (2003b) Spirochete-platelet attachment and thrombocytopenia in murine relapsing fever borreliosis. Blood 102:2843–2850

Alugupalli KR, Leong JM, Woodland RT, Muramatsu M, Honjo T, Gerstein RM (2004) B1b Lymphocytes confer T cell-independent long-lasting immunity. Immunity 21:379–390

Alugupalli KR, Akira S, Lien E, Leong JM (2007) MyD88- and Bruton's tyrosine kinase-mediated signals are essential for T cell-independent pathogen-specific IgM responses. J Immunol 178:3740–3749

Amsbaugh DF, Hansen CT, Prescott B, Stashak PW, Barthold DR, Baker PJ (1972) Genetic control of the antibody response to type 3 pneumococcal polysaccharide in mice. I. Evidence that an X-linked gene plays a decisive role in determining responsiveness. J Exp Med 136:931–949

Andersson J, Sjoberg O, Moller G (1972) Induction of immunoglobulin and antibody synthesis in vitro by lipopolysaccharides. Eur J Immunol 2:349–353

Ansel KM, Harris RB, Cyster JG (2002) CXCL13 is required for B1 cell homing, natural antibody production, and body cavity immunity. Immunity 16:67–76

Arimitsu Y, Akama K (1973) Characterization of protective antibodies produced in mice infected with *Borrelia duttonii*. Jpn J Med Sci Biol 26:229–237

Balasz M, Martin F, Zhou T, Kearney JF (2002) Blood dendritic cells interact with splenic marginal zone B cells to initiate T-independent immune responses. Immunity 17:341–352

Barbour AG (1990) Antigenic variation of a relapsing fever *Borrelia* species. Annu Rev Microbiol 44:155–171

Barbour AG, Bundoc V (2001) In vitro and in vivo neutralization of the relapsing fever agent *Borrelia hermsii* with serotype-specific immunoglobulin M antibodies. Infect Immun 69:1009–1015

Belperron AA, Dailey CM, Bockenstedt LK (2005) Infection-induced marginal zone B cell production of *Borrelia hermsii*-specific antibody is impaired in the absence of CD1d. J Immunol 174:5681–5686

Berland R, Wortis HH (2002) Origins and functions of B-1 cells with notes on the role of CD5. Annu Rev Immunol 20:253–300

Bernasconi NL, Traggiai E, Lanzavecchia A (2002) Maintenance of serological memory by polyclonal activation of human memory B cells. Science 298:2199–2202

Boes M (2000) Role of natural and immune IgM antibodies in immune responses. Mol Immunol 37:1141–1149

Bondada S, Wu H, Robertson DA, Chelvarajan RL (2000) Accessory cell defect in unresponsiveness of neonates and aged to polysaccharide vaccines. Vaccine 19:557–565

Breukels MA, Zandvoort A, Rijkers GT, Lodewijk ME, Klok PA, Harms G, Timens W (2005) Complement dependency of splenic localization of pneumococcal polysaccharide and conjugate vaccines. Scand J Immunol 61:322–328

Briles DE, Nahm M, Schroer K, Davie J, Baker P, Kearney J, Barletta R (1981) Anti-phosphocholine antibodies found in normal mouse serum are protective against intravenous infection with type 3 *Streptococcus pneumoniae*. J Exp Med 153:694–705

Cabatingan MS, Schmidt MR, Sen R, Woodland RT (2002) Naive B lymphocytes undergo homeostatic proliferation in response to B cell deficit. J Immunol 169:6795–6805

Cadavid D, Thomas DD, Crawley R, Barbour AG (1994) Variability of a bacterial surface protein and disease expression in a possible mouse model of systemic Lyme borreliosis. J Exp Med 179:631–642

Carter RH, Fearon DT (1989) Polymeric C3dg primes human B lymphocytes for proliferation induced by anti-IgM. J Immunol 143:1755–1760

Carter RH, Spycher MO, Ng YC, Hoffman R, Fearon DT (1988) Synergistic interaction between complement receptor type 2 and membrane IgM on B lymphocytes. J Immunol 141:457–463

Carvalho TL, Mota-Santos T, Cumano A, Demengeot J, Vieira P (2001) Arrested B lymphopoiesis and persistence of activated B cells in adult interleukin $7^{-/-}$ mice. J Exp Med 194:1141–1150

Chelvarajan RL, Gilbert NL, Bondada S (1998) Neonatal murine B lymphocytes respond to polysaccharide antigens in the presence of IL-1 and IL-6. J Immunol 161:3315–3324

Chelvarajan RL, Raithatha R, Venkataraman C, Kaul R, Han SS, Robertson DA, Bondada S (1999) CpG oligodeoxynucleotides overcome the unresponsiveness of neonatal B cells to stimulation with the thymus-independent stimuli anti-IgM and TNP-Ficoll. Eur J Immunol 29:2808–2818

Chelvarajan RL, Collins SM, Doubinskaia IE, Goes S, Van Willigen J, Flanagan D, De Villiers WJ, Bryson JS, Bondada S (2004) Defective macrophage function in neonates and its impact on unresponsiveness of neonates to polysaccharide antigens. J Leukoc Biol 75:982–994

Chelvarajan RL, Collins SM, Van Willigen JM, Bondada S (2005) The unresponsiveness of aged mice to polysaccharide antigens is a result of a defect in macrophage function. J Leukoc Biol 77:503–512

Chelvarajan RL, Liu Y, Popa D, Getchell ML, Getchell TV, Stromberg AJ, Bondada S (2006) Molecular basis of age-associated cytokine dysregulation in LPS-stimulated macrophages. J Leukoc Biol 79:1314–1327

Chinen J, Shearer WT (2002) Molecular virology and immunology of HIV infection. J Allergy Clin Immunol 110:189–198

Chong Y, Ikematsu H, Kikuchi K, Yamamoto M, Murata M, Nishimura M, Nabeshima S, Kashiwagi S, Hayashi J (2004) Selective CD27$^+$ (memory) B cell reduction and characteristic B cell alteration in drug-naive and HAART-treated HIV type 1-infected patients. AIDS Res Hum Retroviruses 20:219–226

Connolly SE, Benach JL (2001) The spirochetemia of murine relapsing fever is cleared by complement-independent bactericidal antibodies. J Immunol 167:3029–3032

Connolly SE, Thanassi DG, Benach JL (2004) Generation of a complement-independent bactericidal IgM against a relapsing fever *Borrelia*. J Immunol 172:1191–1197

Corley RB, Morehouse EM, Ferguson AR (2005) IgM accelerates affinity maturation. Scand J Immunol 62 [Suppl 1]:55–61

Couderc J, Fevrier M, Duquenne C, Sourbier P, Liacopoulos P (1987) Xid mouse lymphocytes respond to TI-2 antigens when co-stimulated by TI-1 antigens or lymphokines. Immunology 61:71–76

de Vinuesa CG, Cook MC, Ball J, Drew M, Sunners Y, Cascalho M, Wabl M, Klaus GG, MacLennan IC (2000) Germinal centers without T cells. J Exp Med 191:485–494

Dullforce P, Sutton DC, Heath AW (1998) Enhancement of T cell-independent immune responses in vivo by CD40 antibodies. Nat Med 4:88–91

Ellmeier W, Jung S, Sunshine MJ, Hatam F, Xu Y, Baltimore D, Mano H, Littman DR (2000) Severe B cell deficiency in mice lacking the Tec kinase family members Tec and Btk. J Exp Med 192:1611–1624

Fagarasan S, Honjo T (2000) T-Independent immune response: new aspects of B cell biology. Science 290:89–92

Faili A, Aoufouchi S, Gueranger Q, Zober C, Leon A, Bertocci B, Weill JC, Reynaud CA (2002) AID-dependent somatic hypermutation occurs as a DNA single-strand event in the BL2 cell line. Nat Immunol 3:815–821

Fehr T, Naim HY, Bachmann MF, Ochsenbein AF, Spielhofer P, Bucher E, Hengartner H, Billeter MA, Zinkernagel RM (1998) T-cell independent IgM and enduring protective IgG antibodies induced by chimeric measles viruses. Nat Med 4:945–948

Feng SH, Stein KE (1991) VH gene family expression in mice with the *xid* defect. J Exp Med 174:45–51

Fikrig E, Barthold SW, Chen M, Grewal IS, Craft J, Flavell RA (1996) Protective antibodies in murine Lyme disease arise independently of CD40 ligand. J Immunol 157:1–3

Garcia-Monco JC, Miller NS, Backenson PB, Anda P, Benach JL (1997) A mouse model of *Borrelia* meningitis after intradermal injection. J Infect Dis 175:1243–1245

Gebbia JA, Monco JC, Degen JL, Bugge TH, Benach JL (1999) The plasminogen activation system enhances brain and heart invasion in murine relapsing fever borreliosis. J Clin Invest 103:81–87

Guinamard R, Okigaki M, Schlessinger J, Ravetch JV (2000) Absence of marginal zone B cells in Pyk-2-deficient mice defines their role in the humoral response. Nat Immunol 1:31–36

Ha SA, Tsuji M, Suzuki K, Meek B, Yasuda N, Kaisho T, Fagarasan S (2006) Regulation of B1 cell migration by signals through Toll-like receptors. J Exp Med 203:2541–2550

Haas KM, Poe JC, Steeber DA, Tedder TF (2005) B-1a and B-1b cells exhibit distinct developmental requirements and have unique functional roles in innate and adaptive immunity to *S. pneumoniae*. Immunity 23:7–18

Hiernaux JR, Jones JM, Rudbach JA, Rollwagen F, Baker PJ (1983) Antibody response of immunodeficient (xid) CBA/N mice to *Escherichia coli* 0113 lipopolysaccharide, a thymus-independent antigen. J Exp Med 157:1197–1207

Hopken UE, Achtman AH, Kruger K, Lipp M (2004) Distinct and overlapping roles of CXCR5 and CCR7 in B-1 cell homing and early immunity against bacterial pathogens. J Leukoc Biol 76:709–718

Hsu MC, Toellner KM, Vinuesa CG, Maclennan IC (2006) B cell clones that sustain long-term plasmablast growth in T-independent extrafollicular antibody responses. Proc Natl Acad Sci U S A 103:5905–5910

Janeway C, Travers P, Walport M, Shlomchik M (2004) Immunobiology: The immune system in health and disease, 6th edn. Garland Publishing, New York

Jefferies CA, O'Neill LA (2004) Bruton's tyrosine kinase (Btk) – the critical tyrosine kinase in LPS signalling? Immunol Lett 92:15–22

Jefferies CA, Doyle S, Brunner C, Dunne A, Brint E, Wietek C, Walch E, Wirth T, O'Neill LA (2003) Bruton's tyrosine kinase is a Toll/interleukin-1 receptor domain-binding protein that participates in nuclear factor kappaB activation by Toll-like receptor 4. J Biol Chem 278:26258–26264

Kantor AB, Stall AM, Adams S, Watanabe K, Herzenberg LA (1995) De novo development and self-replenishment of B cells. Int Immunol 7:55–68

Kantor AB, Merrill CE, Herzenberg LA, Hillson JL (1997) An unbiased analysis of V(H)-D-J(H) sequences from B-1a B-1b, and conventional B cells. J Immunol 158:1175–1186

Kawai T, Adachi O, Ogawa T, Takeda K, Akira S (1999) Unresponsiveness of MyD88-deficient mice to endotoxin. Immunity 11:115–122

Kelly DF, Moxon ER, Pollard AJ (2004) *Haemophilus influenzae* type b conjugate vaccines. Immunology 113:163–174

Kelly DF, Pollard AJ, Moxon ER (2005) Immunological memory: the role of B cells in long-term protection against invasive bacterial pathogens. JAMA 294:3019–3023

Khan WN, Alt FW, Gerstein RM, Malynn BA, Larsson I, Rathbun G, Davidson L, Muller S, Kantor AB, Herzenberg LA et al (1995) Defective B cell development and function in Btk-deficient mice. Immunity 3:283–299

Khan WN, Nilsson A, Mizoguchi E, Castigli E, Forsell J, Bhan AK, Geha R, Sideras P, Alt FW (1997) Impaired B cell maturation in mice lacking Bruton's tyrosine kinase (Btk) and CD40. Int Immunol 9:395–405

Knoops L, Louahed J, Renauld JC (2004) IL-9-induced expansion of B-1b cells restores numbers but not function of B-1 lymphocytes in xid mice. J Immunol 172:6101–6106

Kruetzmann S, Manuela Rosado M, Weber H, Germing U, Tournilhac O, Peter HH, Berner R, Peters A, Boehm T, Plebani A et al (2003) Human immunoglobulin M memory B cells controlling *Streptococcus pneumoniae* infections are generated in the spleen. J Exp Med 197:939–945

Kurt-Jones EA, Popova L, Kwinn L, Haynes LM, Jones LP, Tripp RA, Walsh EE, Freeman MW, Golenbock DT, Anderson LJ, Finberg RW (2000) Pattern recognition receptors TLR4 and CD14 mediate response to respiratory syncytial virus. Nat Immunol 1:398–401

Landers CD, Chelvarajan RL, Bondada S (2005) The role of B cells and accessory cells in the neonatal response to TI-2 antigens. Immunol Res 31:25–36

Leadbetter EA, Rifkin IR, Hohlbaum AM, Beaudette BC, Shlomchik MJ, Marshak-Rothstein A (2002) Chromatin-IgG complexes activate B cells by dual engagement of IgM and Toll-like receptors. Nature 416:603–607

Lesinski GB, Westerink MA (2001) Novel vaccine strategies to T-independent antigens. J Microbiol Methods 47:135–149

MacLennan IC, Garcia de Vinuesa C, Casamayor-Palleja M (2000) B-cell memory and the persistence of antibody responses. Philos Trans R Soc Lond Biol Sci 355:345–350

Maizels N, Bothwell A (1985) The T-cell-independent immune response to the hapten NP uses a large repertoire of heavy chain genes. Cell 43:715–720

Maizels N, Lau JC, Blier PR, Bothwell A (1988) The T-cell independent antigen NP-Ficoll, primes for a high-affinity IgM anti-NP response. Mol Immunol 25:1277–1282

Malley R, Henneke P, Morse SC, Cieslewicz MJ, Lipsitch M, Thompson CM, Kurt-Jones E, Paton JC, Wessels MR, Golenbock DT (2003) Recognition of pneumolysin by Toll-like receptor 4 confers resistance to pneumococcal infection. Proc Natl Acad Sci U S A 100:1966–1971

Martin F, Kearney JF (2000) B-cell subsets and the mature preimmune repertoire. Marginal zone and B1 B cells as part of a "natural immune memory". Immunol Rev 175:70–79

Martin F, Kearney JF (2001) B1 cells: similarities and differences with other B cell subsets. Curr Opin Immunol 13:195–201

Martin F, Oliver AM, Kearney JF (2001) Marginal zone and B1 B cells unite in the early response against T- independent blood-borne particulate antigens. Immunity 14:617–629

McHeyzer-Williams LJ, Driver DJ, McHeyzer-Williams MG (2001) Germinal center reaction. Curr Opin Hematol 8:52–59

McHeyzer-Williams MG (2003) B cells as effectors. Curr Opin Immunol 15:354–361

Mizuno T, Rothstein TL (2003) Cutting edge: CD40 engagement eliminates the need for Bruton's tyrosine kinase in B cell receptor signaling for NF-kappa B. J Immunol 170:2806–2810

Mizuno T, Rothstein TL (2005) B cell receptor (BCR) cross-talk: CD40 engagement creates an alternate pathway for BCR signaling that activates I kappa B kinase/I kappa B alpha/NF-kappa B without the need for PI3 K and phospholipase C gamma. J Immunol 174:6062–6070

Montecino-Rodriguez E, Dorshkind K (2006) New perspectives in B-1 B cell development and function. Trends Immunol 27:428–433

Montecino-Rodriguez E, Leathers H, Dorshkind K (2006) Identification of a B-1 B cell-specified progenitor. Nat Immunol 7:293–301

Mosier DE, Scher I, Paul WE (1976) In vitro responses of CBA/N mice: spleen cells of mice with an X-linked defect that precludes immune responses to several thymus-independent antigens can respond to TNP-lipopolysaccharide. J Immunol 117:1363–1369

Mosier DE, Zaldivar NM, Goldings E, Mond J, Scher I, Paul WE (1977) Formation of antibody in the newborn mouse: study of T-cell-independent antibody response. J Infect Dis Suppl 136: S14–S19

Muller G, Hopken UE, Lipp M (2003) The impact of CCR7 and CXCR5 on lymphoid organ development and systemic immunity. Immunol Rev 195:117–135

Muramatsu M, Kinoshita S, Fagarasan S, Yamada Y, Shinkai Y, Honjo T (2000) Class switch recombination and hypermutation require Activation-induced cytidine deaminase (AID), a potential RNA editing enzyme. Cell 102:553–563

Nagai Y, Akashi S, Nagafuku M, Ogata M, Iwakura Y, Akira S, Kitamura T, Kosugi A, Kimoto M, Miyake K (2002) Essential role of MD-2 in LPS responsiveness and TLR4 distribution. Nat Immunol 3:667–672

Obukhanych TV, Nussenzweig MC (2006) T-independent type II immune responses generate memory B cells. J Exp Med 203:305–310

Ochsenbein AF, Pinschewer DD, Odermatt B, Carroll MC, Hengartner H, Zinkernagel RM (1999) Protective T cell-independent antiviral antibody responses are dependent on complement. J Exp Med 190:1165–1174

Pinschewer DD, Ochsenbein AF, Satterthwaite AB, Witte ON, Hengartner H, Zinkernagel RM (1999) A Btk transgene restores the antiviral TI-2 antibody responses of xid mice in a dose-dependent fashion. Eur J Immunol 29:2981–2987

Satterthwaite AB, Cheroutre H, Khan WN, Sideras P, Witte ON (1997) Btk dosage determines sensitivity to B cell antigen receptor cross-linking. Proc Natl Acad Sci U S A 94:13152–13157

Scher I, Steinberg AD, Berning AK, Paul WE (1975) X-linked B-lymphocyte immune defect in CBA/N mice. II. Studies of the mechanisms underlying the immune defect. J Exp Med 142:637–650

Schnare M, Barton GM, Holt AC, Takeda K, Akira S, Medzhitov R (2001) Toll-like receptors control activation of adaptive immune responses. Nat Immunol 2:947–950

Schroder NW, Opitz B, Lamping N, Michelsen KS, Zahringer U, Gobel UB, Schumann RR (2000) Involvement of lipopolysaccharide binding protein CD14, and Toll-like receptors in the initiation of innate immune responses by Treponema glycolipids. J Immunol 165:2683–2693

Scorpio A, Chabot DJ, Day WA, O'Brien D, K, Vietri NJ, Itoh Y, Mohamadzadeh M, Friedlander AM (2007) Poly-gamma-glutamate capsule-degrading enzyme treatment

enhances phagocytosis and killing of encapsulated *Bacillus anthracis*. Antimicrob Agents Chemother 51:215–222
Selinka HC, Bosing-Schneider R (1988) Xid mice fail to express an anti-dextran immune response but carry alpha(1–3)dextran-specific lymphocytes in their potential repertoire. Eur J Immunol 18:1727–1732
Shang ES, Skare JT, Exner MM, Blanco DR, Kagan BL, Miller JN, Lovett MA (1998) Isolation and characterization of the outer membrane of *Borrelia hermsii*. Infect Immun 66:1082–1091
Snapper CM, Rosas FR, Moorman MA, Mond JJ (1997) Restoration of T cell-independent type 2 induction of Ig secretion by neonatal B cells in vitro. J Immunol 158:2731–2735
Southern PM Jr, Sanford JP (1969) Relapsing fever – a clinical and microbiological review. Medicine 48:129–149
Stall AM, Adams S, Herzenberg LA, Kantor AB (1992) Characteristics and development of the murine B-1b (Ly-1 B sister) cell population. Ann N Y Acad Sci 651:33–43
Szomolanyi-Tsuda E, Welsh RM (1998) T-cell-independent antiviral antibody responses. Curr Opin Immunol 10:431–435
Szomolanyi-Tsuda E, Brien JD, Dorgan JE, Garcea RL, Woodland RT, Welsh RM (2001) Antiviral T-cell-independent type 2 antibody responses induced in vivo in the absence of T and NK cells. Virology 280:160–168
Szomolanyi-Tsuda E, Seedhom MO, Carroll MC, Garcea RL (2006) T cell-independent and T cell-dependent immunoglobulin G responses to polyomavirus infection are impaired in complement receptor 2-deficient mice. Virology 352:52–60
Takeda K, Akira S (2005) Toll-like receptors in innate immunity. Int Immunol 17:1–14
Tarlinton D (2006) B-cell memory: are subsets necessary? Nat Rev Immunol 6:785–790
Thomas JD, Sideras P, Smith CIE, Vorechovsky I, Chapman V, Paul WE (1993) Colocalization of X-linked agammaglobulinemia and X-linked Immunodeficiency genes. Science 261:355–358
Toellner KM, Jenkinson WE, Taylor DR, Khan M, Sze DM, Sansom DM, Vinuesa CG, MacLennan IC (2002) Low-level hypermutation in T cell-independent germinal centers compared with high mutation rates associated with T cell-dependent germinal centers. J Exp Med 195:383–389
Tornberg UC, Holmberg D (1995) B-1a B-1b and B-2 B cells display unique VHDJH repertoires formed at different stages of ontogeny and under different selection pressures. EMBO J 14:1680–1689
Toyama H, Okada S, Hatano M, Takahashi Y, Takeda N, Ichii H, Takemori T, Kuroda Y, Tokuhisa T (2002) Memory B cells without somatic hypermutation are generated from Bcl6-deficient B cells. Immunity 17:329–339
Tung JW, Mrazek MD, Yang Y, Herzenberg LA, Herzenberg LA (2006) Phenotypically distinct B cell development pathways map to the three B cell lineages in the mouse. Proc Natl Acad Sci U S A 103:6293–6298
Vink A, Warnier G, Brombacher F, Renauld JC (1999) Interleukin 9-induced in vivo expansion of the B-1 lymphocyte population. J Exp Med 189:1413–1423
Vinuesa CG, Sunners Y, Pongracz J, Ball J, Toellner KM, Taylor D, MacLennan IC, Cook MC (2001) Tracking the response of Xid B cells in vivo: TI-2 antigen induces migration and proliferation but Btk is essential for terminal differentiation. Eur J Immunol 31:1340–1350
Vos Q, Lees A, Wu Z, Snapper CM, Mond JJ (2000) B-cell activation by T-cell-independent type 2 antigens as an integral part of the humoral immune response to pathogenic microorganisms. Immunol Rev 176:154–170
Wang TT, Lucas AH (2004) The capsule of *Bacillus anthracis* behaves as a thymus-independent type 2 antigen. Infect Immun 72:5460–5463
Watanabe N, Ikuta K, Fagarasan S, Yazumi S, Chiba T, Honjo T (2000) Migration and differentiation of autoreactive B-1 cells induced by activated gamma/delta T cells in anti-erythrocyte immunoglobulin transgenic mice. J Exp Med 192:1577–1586
Weller S, Faili A, Garcia C, Braun MC, Le Deist FF, de Saint Basile GG, Hermine O, Fischer A, Reynaud CA, Weill JC (2001) CD40-CD40L independent Ig gene hypermutation suggests

a second B cell diversification pathway in humans. Proc Natl Acad Sci U S A 98:1166–1170

Woodland RT, Schmidt MR (2005) Homeostatic proliferation of B cells. Semin Immunol 17:209–217

Woodland RT, Schmidt MR, Korsmeyer SJ, Gravel KA (1996) Regulation of B cell survival in xid mice by the proto-oncogene bcl-2. J Immunol 156:2143–2154

Yan SR, Byers DM, Bortolussi R (2004a) Role of protein tyrosine kinase p53/56lyn in diminished lipopolysaccharide priming of formylmethionylleucyl-phenylalanine-induced superoxide production in human newborn neutrophils. Infect Immun 72:6455–6462

Yan SR, Qing G, Byers DM, Stadnyk AW, Al-Hertani W, Bortolussi R (2004b) Role of MyD88 in diminished tumor necrosis factor alpha production by newborn mononuclear cells in response to lipopolysaccharide. Infect Immun 72:1223–1229

Yokota M, Morshed MG, Nakazawa T, Konishi H (1997) Protective activity of *Borrelia duttonii*-specific immunoglobulin subclasses in mice. J Med Microbiol 46:675–680

Yu PW, Tabuchi RS, Kato RM, Astrakhan A, Humblet-Baron S, Kipp K, Chae K, Ellmeier W, Witte ON, Rawlings DJ (2004) Sustained correction of B-cell development and function in a murine model of X-linked agammaglobulinemia (XLA) using retroviral-mediated gene transfer. Blood 104:1281–1290

Zandvoort A, Timens W (2002) The dual function of the splenic marginal zone: essential for initiation of anti-TI-2 responses but also vital in the general first-line defense against blood-borne antigens. Clin Exp Immunol 130:4–11

Zandvoort A, Lodewijk ME, de Boer NK, Dammers PM, Kroese FG, Timens W (2001) CD27 expression in the human splenic marginal zone: the infant marginal zone is populated by naive B cells. Tissue Antigens 58:234–242

Zinkernagel RM (2000) What is missing in immunology to understand immunity? Nat Immunol 1:181–185

Secretory Immunity Following Mutans Streptococcal Infection or Immunization

D. J. Smith(✉) and R. O. Mattos-Graner

1	Introduction	132
2	Mutans Streptococci	133
	2.1 Cariogenicity of Mutans Streptococci	133
	2.2 Initial Colonization	134
3	Molecular Pathogenesis of Disease	134
4	Immune Responses to Mutans Streptococcal Infection in Adults	136
5	Ontogeny of Salivary/Secretory Immunity in Children	138
	5.1 Induction of Mucosal Immunity	138
	2.2 Ontogeny of Mucosal Immunity	138
6	Salivary IgA Antibody Following Initial Mutans Streptococcal Exposure	142
7	Active Immunization	144
	7.1 Targeting Adherence	144
	7.2 Targeting Accumulation	145
8	Methods of Antigen Delivery and Immune Enhancement	146
	8.1 Routes	146
	8.2 Adjuvants	147
	8.3 Delivery Systems	148
9	Passive Immunization	148
10	Summary	151
	References	152

Abstract Salivary IgA antibody responses to mutans streptococci can be observed in early childhood, sometimes even before permanent colonization of the oral biofilm occurs. Many of these early immune responses are directed to components thought to be essential for establishment and emergence of mutans streptococci in the oral biofilm. Initial responses are likely to be modulated by antigen dose, by immunological maturity, and by previous encounters with similar antigenic epitopes in the pioneer commensal flora. Our understanding of these modulating

D. J. Smith
Department of Immunology, The Forsyth Institute, 140 The Fenway, Boston, MA 02115, USA
e-mail: dsmith@forsyth.org

factors is modest and is an opportunity for continued investigation. Under controlled conditions of infection, experimental vaccine approaches have repeatedly shown that infection and disease can be modified in the presence of elevated levels of antibody in the oral cavity. Protection can be observed regardless of antibody isotype or method used to actively or passively provide the immune reagent. Limited clinical trials have supported the utility of both of these approaches in humans. Refinements in antigen formulation, delivery vehicles, enhancing agents and routes of application, coupled with approaches that are timed to intercept most vulnerable periods of infection of primary and permanent dentition may well provide the healthcare practitioner with an additional tool to maintain oral health.

Abbreviations MS: Mutans streptococci; *S. mutans*: *Streptococcus mutans*; Gbp: Glucan-binding protein; Ag I/II: Antigen I/II; *S. sobrinus*: *Streptococcus sobrinus*; SpaA: Surface protein antigen A; WapA: Cell wall-associated protein; Gtf: Glucosyltransferase; CA-Gtf: Cell-associated glucosyltransferase; CF-Gtf: Culture supernatant glucosyltransferase; GCF: Gingival crevicular fluid; IgG: Immunoglobulin G; IgA: Immunoglobulin A; IgM: Immunoglobulin M; IgY: Immunoglobulin Y; SIgA: Secretory immunoglobulin A; MALT: Mucosa-associated lymphoid tissue; GALT: Gut-associated lymphoid tissue; T cells: Thymus-derived lymphocytes; B cells: Bone marrow-derived lymphocytes; MHC: Major histocompatibility complex; *S. mitis*: *Streptococcus mitis*; *S. salivarius*: *Streptococcus salivarius*; ELISA: Enzyme-linked immunosorbent assay; NALT: Nasal-associated lymphoid tissue; SBR-CT: Salivary binding reception-cholera toxin; scFc: Single chain Fc; VHH: Variable domain of llama heavy chain homodimeric IgG

1 Introduction

Oral biofilms are exposed to a variety of innate and adaptive host components which enter via the major and minor salivary glands, gingival crevice, or perfuse through the oral epithelium. The impact, both short and long-term, of this exposure on individual members of the biofilm is difficult to measure in the native state. Adaptive host responses occur to permanent constituents as well as to transient bacteria, provided microorganisms are in sufficient concentration or duration of exposure. The level of colonization need not be high since distinct and complex salivary immune responses can be observed to pioneer microbiota within a few weeks of life. Oral immune responses are complex with regard to specificity, amount, isotype, and source. These responses experience a degree of modulation as microorganisms adapt to a commensal state through mechanisms that are not well understood. Oral bacteria regarded as pathogens under certain conditions are frequently members of the normal oral flora. Hallmarks of host responses to these pathogens can be identified before and during diseases caused by them.

Against this backdrop, there has been significant effort to identify and understand host responses to one of these pathogens, *Streptococcus mutans*, and related species (mutans streptococci) which are considered to initiate dental caries, especially in children. Given the presumption that induction of host responses to bacterial infections can be a means to interfere with their pathogenic consequences, many investigators have adopted vaccine approaches in order to block the entry or reentry of mutans streptococci into established oral biofilms and thus to thwart the dental caries disease process. This review will briefly outline the relationship of mutans streptococci with dental caries and highlight those features of most interest immunologically. The characteristics of immune responses to mutans streptococcal exposure will be described, together with the challenges of interpreting the dynamic relationships of host response, infection, and disease. Adaptive and passive approaches that seek to modify the colonization or disease potential of these cariogenic streptococci will also be discussed.

2 Mutans Streptococci

2.1 Cariogenicity of Mutans Streptococci

Dental caries is an infectious disease that occurs because of an imbalance in the homeostasis between the host and oral flora. This imbalance promotes the emergence of cariogenic microorganisms in the oral (dental) biofilm. Cariogenic organisms are defined as those able to cause tooth demineralization because of their capacity to accumulate, produce, and tolerate extremely low pHs in the dental plaque. Abundant evidence exists implicating mutans streptococci as the principal organisms associated with the onset and progression of dental caries (see Loesche 1986; Tanzer 1995, for reviews). Although mutans streptococci include seven distinct species, only two, *Streptococcus mutans* and *Streptococcus sobrinus,* are exclusively isolated from humans (Whiley and Beighton 1998). These streptococci can accumulate in the plaque enough to cause a decrease in biofilm pH to less than 5 by virtue of synthesis of lactic acid from the fermentative metabolism of dietary sugars and from intracellular and extracellular reservoirs of saccharides. The low pH responsible for tooth demineralization is tolerated by mutans streptococci that can physiologically adapt to this acidic condition, thus promoting irreversible tooth damage.

Mutans streptococci (MS) have been identified worldwide, although the frequency and intensity of infection and associations with dental caries varied with the population studied (Tanzer et al. 2001). An early association between childhood caries and mutans streptococcal infection was made by Alaluusua and Renkonen (1983) who reported that children with detectable levels of MS at 2 years of age had a caries incidence 11-fold higher than children not colonized at the same age. Later, Köhler et al. (1988) observed that children who had been colonized by *S. mutans* before 2 years of age had caries scores which were many-fold higher when they were 4 years old, compared with children who were colonized later.

Subsequent prospective studies have confirmed that preventive programs which significantly reduce MS levels of mothers, delayed colonization of their children, which had a significant long-lasting impact on their caries development (Kohler and Andreen 1994; Kohler et al. 1983).

2.2 Initial Colonization

Mutans streptococci require nonshedding tooth surfaces to become permanently established in the oral cavity (Catalanotto et al. 1975), although more sensitive DNA-specific probes suggest that these microorganisms may be found in the oral cavity prior to tooth eruption (Tanner et al. 2002). These infections are primarily transmitted vertically from the primary caregiver (Li and Caufield 1995; Klein et al. 2004), though there is evidence for horizontal transmission (Mattos-Graner et al. 2001b). Although children are repeatedly exposed to maternally derived mutans streptococci, permanent colonization of the primary teeth typically occurs in the 2nd and 3rd year of life. The natural history of MS colonization was analyzed in a prospective study of 46 children from birth to 5 years of age whose mothers had relatively high levels of MS (Caufield et al. 1993). They observed that all children who acquired MS were colonized between the ages of 19 and 31 months. The reasons that account for this so-called window of infectivity are not fully understood but may be associated with the eruption of molars, which provide virgin and retentive occlusal surfaces for colonization. Later studies performed in American children confirmed the notion of a window of infectivity (Smith et al. 1998). Interestingly, both studies described a subset of children who remained free of mutans streptococci in the biofilm until much later in childhood. It was hypothesized that if the biofilm matured on the fully erupted primary dentition without mutans streptococci, then an appropriate niche would not become available to these microorganisms until permanent teeth appeared.

Conditions exist whereby the window of infectivity can open earlier in life. If children are under high infectious challenge as a result of frequent exposure to heavily infected caregivers, coupled with excessive sucrose intake, then infection may be demonstrated within the 1st year (Kohler and Andreen 1994; Tanner et al. 2002; Mattos-Graner et al. 2001a; Karn et al. 1998). Other host and environmental factors may influence the time of MS acquisition, one of which may be the status of immunological maturation. Whether the age of infection is early or late, once mutans streptococci become members of the oral biofilm, they remain permanent residents.

3 Molecular Pathogenesis of Disease

Mutans streptococci produce several proteins important to the pathogenesis of dental caries. These proteins are involved in the ability to colonize teeth and to impose their ecological advantage over other commensal organisms when the dental biofilm is under environmental stress. For initial colonization, MS are able to adhere to saliva-coated tooth surfaces by specific interactions (adhesins) with salivary components and with surface proteins of other bacterial species. Following adhesion, MS are able to increase in proportion to other microorganisms within the biofilm community through the enzymatic synthesis (via glucosyltransferases) of an extracellular matrix of water-insoluble glucans, which then interact with glucan binding proteins (Gbp) on bacterial surfaces (see Banas and Vickerman 2003). Expression of these and other virulence factors are under environmental regulation via intra- and interspecies communication systems (Senadherra et al. 2005, 2007), which allow these organisms to increase in proportion when under extreme stress conditions that can result from external environmental factors and from their own bacterial products (Burne et al. 1997; Quivey et al. 2001). Because of their ability to adapt to acidic environments, mutans streptococci can accumulate during periods of biofilm acidification, thus increasing the extent of pH drop, resulting in tooth demineralization. Understanding the molecular mechanisms of pathogenesis has allowed investigators to identify targets for controlling *S. mutans* infection and virulence.

The group of surface adhesins expressed by *S. mutans*, first discovered by Russell and Lehner (1978), are variously termed antigen I/II (Ag I/II), SpaP, Pac or P1, and in *S. sobrinus,* SpaA. The cell wall-anchored *S. mutans* adhesin binds to a high-molecular-weight glycoprotein called salivary agglutinin, or gp340. Antigen I/II has been reported to interact with the gp340 ligand via an alanine repeat region in an N-terminal domain and by conformational epitopes between the alanine-rich and a more centrally located proline-rich region (Seifert et al. 2004). Polypeptides from this family of proteins have also been observed in other commensal species such as *Streptococcus gordonii* and *Streptococcus intermedius* (reviewed in Jenkinson and Lamont 2005). *S. mutans* also expresses a cell wall-associated protein (WapA), which was suggested to bind saliva-coated smooth surfaces and to participate in sucrose-dependent adherence, but its role in cariogenesis is not fully understood (Harrington and Russell 1993; Qian and Dao 1993).

Glucosyltransferases (Gtf) have a crucial role in the accumulation of *S. mutans* in the biofilm, since they catalyze the synthesis of an extracellular matrix of glucans from sucrose (Smith 2002). *S. mutans* expresses three Gtf isotypes (GtfB, GtfC, and GtfD) that produce glucans with distinct degrees of water solubility, depending of the proportion and type of glycosidic linkages. The isotypes GtfB and GtfC synthesize $\alpha(1-3)$-rich water-insoluble glucans, while GtfD synthesizes water-soluble glucans that are rich in $\alpha(1-6)$ glycosidic linkages. *S. sobrinus* also expresses several Gtfs with significant homologies and water solubility of glucan products to those of *S. mutans*. Gtfs have two functional domains, a catalytic domain located in the N-terminal half of the molecule, and a glucan-binding domain located in the carboxy-terminal third. The glucan-binding domain contains amino acid repeats that resemble ligand-binding domains of various Gram-positive bacteria and that were also found to be necessary for the catalytic activity. The number of repeats in this domain affects the structure of glucans, and thus may affect adherence properties (Banas and Vickerman 2003).

The molecular interaction of *S. mutans* cells with the extracellular glucan matrix is not fully understood, but a heterogeneous group of proteins have been shown to bind glucans, and for that property were named glucan-binding proteins (Gbps).

S. mutans expresses at least four distinct Gbps (GpbA, GbpB, GbpC, and GbpD) whose letter description is based on the order in which they were first described. Apart from their affinity for glucan, genetic and biochemical studies of each protein have indicated a variety of biological functions (reviewed by Banas and Vickerman 2003). GbpA has homology with the glucan-binding domain of *S. mutans* GtfB and GtfC (Banas et al. 1990). In contrast, GbpB shows no homology with other glucan-binding proteins or domains, but may function in the maintenance of cell wall integrity (Mattos-Graner et al. 2001a, 2006; Chia et al. 2001). Production of GbpB (Smith et al. 1994) appears to promote the ability of *S. mutans* genotypes to grow in artificial biofilms (Mattos-Graner et al. 2001a), but mechanisms regulating expression of this protein are still unclear (Chia et al. 2001; Mattos-Graner et al. 2006; Senadheera et al. 2005). GbpC is expressed in conditions of osmotic stress, allowing involvement in *S. mutans* aggregation in the presence of glucan and dextran and adherence of *S. mutans* to saliva-coated tooth surfaces (Sato et al. 1997). GbpD has characteristics of both dextran-binding and lipase activity (Shah and Russell 2004). *S. sobrinus* also secretes several glucan-binding proteins (Smith et al. 1998).

4 Immune Responses to Mutans Streptococcal Infection in Adults

Antibody activity to mutans streptococcal antigens can be detected in both salivas and sera (reviewed in Michalek and Childers 1990). These antibody specificities are not unexpected, given the life-long presence of mutans streptococci in the oral biofilm. The presence of IgG antibody to mutans streptococci can be detected in infant sera as a consequence of placental transfer (Luo et al. 1988) and reflects maternal experience with these microorganisms. Children then begin to synthesize serum antibody to mutans streptococcal antigens in early childhood. Serum IgG antibody levels increase during childhood and remain detectable throughout life. Most studies that examined the association between serum antibody to *S. mutans* and dental disease were conducted in young adults and usually found negative correlations between IgG antibody and disease levels (Block et al. 1979; Challacombe 1980; Gregory et al. 1986). In contrast, studies of older adults revealed a positive correlation between serum IgG antibody to these cariogenic streptococci and their cumulative dental caries experience (Kent et al. 1992).

The local oral expression of the systemic immune response can be studied, in part, in the gingival crevicular fluid (GCF). The antibody component of this fluid is a mixture of serum-derived immunoglobulins and locally synthesized IgG and IgA. As in serum, IgG is the dominant isotype. Several studies have shown that at least some of the IgG antibody to *S. mutans* in GCF is locally derived and that antibody levels can vary from site to site in the same individual (Lehner et al. 1987; Camling et al. 1992; Smith et al. 1994). Attempts to draw general conclusions about the significance of naturally synthesized GCF antibody on colonization and disease are complicated by differences in study designs and the intricate dynamics of biofilm maintenance. For example, Camling and co-workers (1991) found no correlation between GCF antibody activity and levels of indigenous *S. mutans* colonizing buccal surfaces of the first permanent molars in a group of young school-age children. In contrast, studies in older adults suggested that diminished recolonization of cleaned tooth surfaces with indigenous *S. mutans* was associated with the presence of IgG antibody in the locally secreted GCF (Smith et al. 1994).

Attempts to draw conclusions on the relationship between the levels or specificity of naturally formed salivary IgA antibody and dental disease in adults have been no less problematic. Some investigators have found that elevated levels of IgA antibody to intact bacteria reflected lower caries experience, while others found consistent elevations in antibody levels only when active carious lesions were present (reviewed in Michalek and Childers 1990; Smith and Taubman 1991). A higher diversity of salivary IgA-reactive antigens of *S. mutans* was described for caries-free, when compared to caries-active 12- to 13-year-old Thai children (Bratthall et al. 1997). Again, comparisons of saliva studies are challenging for some of the same reasons as indicated for systemic comparisons. In addition, salivary studies are further complicated by factors such as dietary intake of fermentable carbohydrates, salivary flow rates, the absorptive capacity of microbiota, and the contribution of GCF antibody when whole saliva is used for analysis. What is clear is that salivary IgA antibody to a variety of mutans streptococcal components is present to some extent in parotid, submandibular/sublingual, and minor gland fluids of most healthy adults. After mucosal immune maturation is complete (see Sect. 5) parotid IgA (Smith et al. 1992; Challacombe et al. 1995) and parotid IgA antibody levels to *S. mutans* (Percival et al. 1997) remain relatively constant throughout life, although IgA concentrations in stimulated minor gland saliva have been reported to show a significant reduction with age (Smith et al. 1992).

The relationships between mutans streptococcal levels or dental caries and host response in these cross-sectional snapshots of older children and adults must be understood in context. Subjects have had life-long associations with mutans streptococci. During this time, immune responses to these cariogenic members of the normal oral biofilm may have several, overlapping effects. For example, initial immune responses to mutans streptococcal antigens may influence the time and rate at which these streptococci join the biofilms of the primary dentition. The level and specificity of the immune response may also modify the ability of commensal mutans streptococci to colonize subsequently erupting primary or permanent teeth. Antibody from any of the sources discussed above may influence the accumulation of a cariogenic flora at various stages of infection. However, dental disease may progress under diverse environmental challenges in spite of the putatively protective immune response. This disease would likely increase the antigenic load and result in additional responses. Thus the levels of oral immune parameters, infection, and past or present disease in adulthood may tell us little about the effect of the immune response on the establishment or pathogenesis of oral microorganisms. Perhaps examination of relationships between bacterial colonization and immune responses that take place during the initial formation of the oral biofilm may be more instructive.

5 Ontogeny of Salivary/Secretory Immunity in Children

5.1 Induction of Mucosal Immunity

Secretory IgA (SIgA) antibody in the saliva appears as a consequence of induction of immune responses in one or more specialized anatomical areas, collectively known as mucosa-associated lymphoid tissue (MALT) (Brandtzaeg 2007). Swallowed bacteria can be taken up by specialized epithelial M cells overlying Peyer's patches in the lower ileum. These cells then pass bacterial components to dendritic cells, macrophages and B cells. These cells then present antigen to $CD4^+$ helper T cells which, when activated, liberate cytokines which cause mucosal B cells to differentiate, primarily to IgA-committed cells, which then migrate through the lymphatic circulation to the mesenteric lymph nodes, then reach the peripheral circulation through the thoracic duct. These primed mucosal memory/effector B cells home to mucosal sites via mechanisms which include adhesion molecules, chemokines and, in some cases, topically available antigen (Brandtzaeg et al. 2005). IgA dimers then bind poly- immunoglobulin receptor (pIgR) on the glandular epithelium, are endocytosed, and finally secreted into the glandular lumen as secretory IgA (SIgA) to which modified pIgR (aka secretory component) is covalently attached. Once seeded in local mucosal tissue, final differentiation to IgA dimer-secreting plasma cells takes place. Because of the significant regionalization of the mucosal immune system, inductive sites in Waldeyer's ring, which contain MHC class II molecule-expressing follicle-associated epithelium (Johansen et al. 2005), are likely to be important for priming effector B cells to oral antigens and subsequent antibody-secreting plasma cells in the salivary glands. Salivary SIgA antibody levels may also be amplified by the transport of oral antigens via dendritic cells to cervical lymph nodes, which also are thought to have mucosal immune inductive properties.

5.2 Ontogeny of Mucosal Immunity

Effective pediatric dental caries vaccine approaches depend, in part, on an understanding of mucosal immune ontogeny. Ideally, immune intervention would take place during initial mutans streptococcal colonization events. Although some immunological issues which are important for the application of a childhood dental caries vaccine are incompletely resolved, sufficient data are available to formulate reasonable strategies. For example, the natural history of mutans streptococcal infection described above indicates that a child's mucosal immune system needs to be sufficiently mature by the end of the 1st year of life in order for a vaccine to induce an effective response. Since children of this age do not now receive mucosal immunizations, knowledge of the immunological responsiveness of the secretory immune system primarily comes from analysis of mucosal responses to infections with indigenous organisms.

Of necessity, the oral immune environment undergoes rapid, early development. Although secretory IgA antibody in saliva and other secretions is essentially absent at birth (Haworth and Dilling 1966), IgM and IgA-containing immunocytes are present in salivary tissue at this time (Thrane et al. 1987, 1990; Iwase et al. 1987) and salivary epithelial cells contain a secretory component. Following birth, exposure to bacterial, viral, and food antigens causes a rapid expansion of IgA plasma cells in mucosal lamina propria (Spenser et al. 1990). Immunomodulatory factors in breast milk apparently enhance this expansion since neonates receiving breast milk have significantly higher concentrations of salivary IgA than those receiving nutrition by other means (Hayes et al. 1999; Newburg and Walker 2007). Salivary (secretory) IgA concentrations continue to increase at a higher rate in breast vs bottle fed infants during the first 6 months of life (Fitzsimmons et al. 1994). Salivary IgA concentrations of 30 µg/ml are typical at this age. Children between 12 and 24 months exhibit a broader range of generally higher salivary IgA concentrations (median concentrations of 50–60 µg/ml, although some range to greater than 200 µg/ml) than seen in the 1st year (Gahnberg et al. 1985).

Mature SIgA, i.e., dimeric IgA with bound secretory component, is the principal salivary immunoglobulin secreted by 1 month of age (Smith et al. 1989). At this early stage, many infant salivas are dominated by the IgA1 isotype (Smith et al. 1989; Fitzsimmons et al. 1994; Childers et al. 2003). Many oral biofilms of infants contain relatively large proportions of bacteria that secrete IgA1 proteases, a condition that would theoretically reduce the protective capacity of salivary IgA antibody at this time. However, there is little evidence of proteolytic cleavage of IgA in whole salivas from infants with IgA1 protease-secreting flora (Smith et al. 1998), suggesting that potential cleavage is principally occurring within the biofilm, rather than in the planktonic phase. After 6 months of age, most children exhibit a more adult-like distribution of salivary IgA1 and IgA2 subclasses, both of which contain antibody specificities to oral commensal microbiota (Smith and Taubman 1992; Cole et al. 1998; R.D. Nogueira et al., unpublished observations).

Commensal flora represent a significant antigenic stimulus to the developing mucosal immune system. As a result, mucosal IgA antibody to pioneer gut (*Escherichia coli*, Gleeson et al. 1985; Mellander et al. 1984) and oral (*Streptococcus mitis* and *S. salivarius*, Smith et al. 1990, 1992; Cole et al. 1999) microbiota can be detected within weeks of initial exposure. Neonatal infections with rotovirus (Jayashree et al. 1988) or poliovirus (Hanson et al. 1987) also induce detectable salivary IgA antibody. IgA antibody to tetanus toxoid and poliovirus appears in the saliva within the first months of life as a consequence of pediatric immunization with these injected and oral vaccines (Smith et al. 1990). The significance of these responses to antigens present in the developing commensal flora is not yet understood. One of the difficulties in quantitatively assessing mucosal immune responses in young children is that, in addition to the increasing concentrations of IgA, antibody reactivity is measured in whole saliva, rather than pure glandular secretions. Also, the increasing diversity of commensal flora and dietary components, eruption of primary dentition, and variation in salivary flow rates have considerable impact on salivary antibody concentration.

Nonetheless, the characteristics of response patterns to initially colonizing streptococci have been followed through the first years of life. Cole and co-workers (1999) measured salivary IgA antibody to formalin-killed *S. mitis* biovar 1, *S. oralis*, *S. mutans*, and *Enterococcus faecalis* in children from birth to 2 years of age. They found no differences in concentrations of IgA antibody to formalin-killed cells of each of these species during this time period. In contrast, our study of the relationship between initial colonization with so-called pioneer streptococci and the corresponding longitudinal development of salivary IgA antibody revealed a different picture (Smith and Taubman 1992). Using culture supernatants from recently isolated *S. mitis* as antigen, Western blots (Fig. 1) were done with two children's salivas taken between 9 to 43 months of age. As the age of the child increased, so did the number and intensity of responses to *S. mitis* components reacting with salivary IgA. A similar finding was observed among 6- to 18-month-old children using boiled whole *S. mitis* cells as antigen (Nogueira et al. 2005). These latter studies suggested that a quantitative increase in the amount of salivary IgA antibody to many *S. mitis* components took place, although this was likely to be influenced by underlying increases in salivary IgA secretion.

Cole and co-workers (1999) also found that much of the salivary IgA antibody reactive with *S. mitis* and *S. salivarius* formalin-killed cells could be removed by

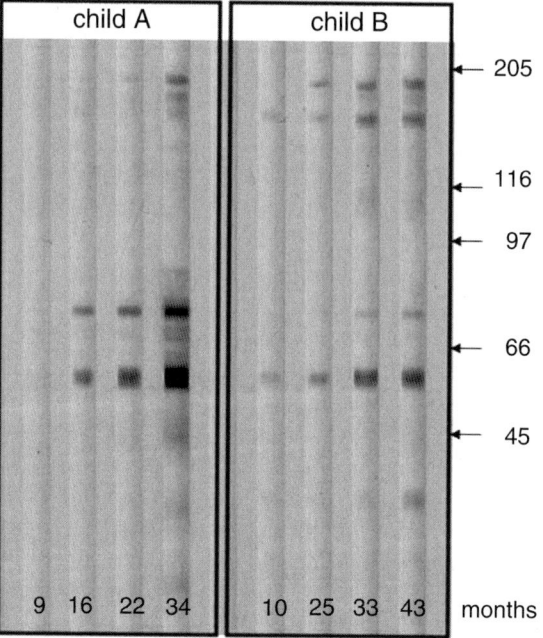

Fig. 1 Western blot of *Streptococcus mitis* culture supernatant component reaction with IgA antibody in salivas from two children. Salivas were collected at 9, 16, 22, and 34 months of life from child A, and 10, 25, 33 and 43 months of life from child B. Migration of standards is indicated on the right edge of the blot by *arrows* and the corresponding kDa

adsorption with other viridans streptococci, including *S. mutans,* which do not colonize until much later. They suggested, therefore, that the predominant salivary IgA antibody response in young children to the four tested streptococcal species was directed to cross-reactive antigens. This may reflect innate-like polyclonal responses observed to commensal gut flora, which are characterized by limited repertoire, cross-reactive SIgA antibody of low affinity (Bouvet and Fischetti 1999; Macpherson et al. 2000). Superimposed on these low-level responses are adaptive, higher-affinity responses to species-specific antigens, which may result from initial bacterial challenge or pathogen exposure. For example, salivas from a separate longitudinally studied group of children were exposed in Western blot to culture supernatants from recently isolated *S. mitis* and *S. mutans* (Smith et al. 1998) Little or no salivary IgA antibody reactivity was observed to *S. mutans* antigens in the 2nd year of life (prior to colonization with mutans streptococci), although the same salivas reacted vigorously with several *S. mitis* components (Fig. 2). These studies would indicate that a significant level of specificity exists in the early mucosal responses to the current and future members of the mature biofilm, superimposed on an underlying modulated response to shared antigens. The challenge is to understand the role which these responses play in the development of biofilm communities. Taken together, the evidence suggests that mucosal immune maturation has developed to the stage at which a detectable salivary IgA immune response to mucosally applied dental caries vaccines could be expected by the end of the 1st year of life, at least to protein antigens.

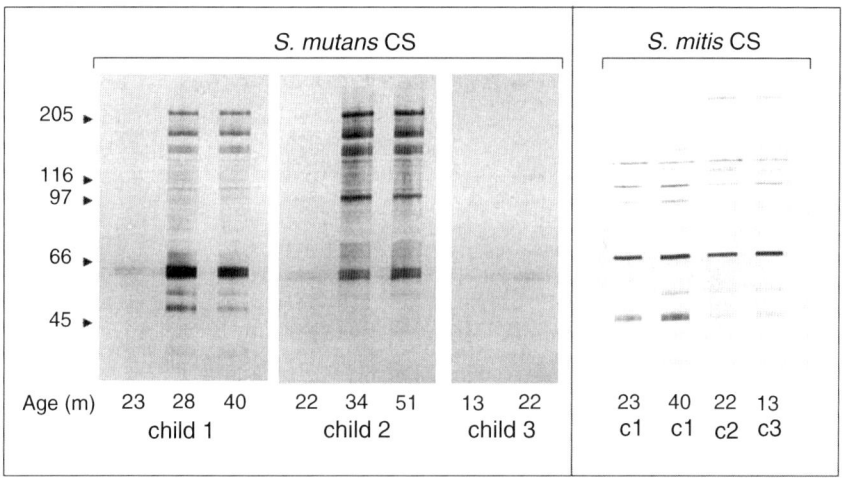

Fig. 2 Western blot of *Streptococcus mutans* (*left side*) and *Streptococcus mitis* (*right side*) culture supernatant (CS) component reactions with IgA antibody in salivas from three children (c). The age at which salivas were collected from each child is indicated below the respective lane. Migration of standards is indicated on the left edge of the blot by *arrows* and the corresponding kDa

6 Salivary IgA Antibody Following Initial Mutans Streptococcal Exposure

Natural exposure to mutans streptococci results in a mucosal immune response to these organisms. This response is often first observed beginning late in the 2nd year of life, when mutans streptococci begin to accumulate on primary tooth surfaces. Western blot and ELISA analyses reveal that the major responses often appear to be directed to streptococcal components which are considered to be important in colonization and accumulation (Smith et al. 1998). Prominent among the components reacting with IgA antibody are antigen I/II, glucosyltransferase, and glucan-binding protein B. Interestingly, salivary IgA reactivity with an unidentified component migrating to 110 kDa is frequently observed in salivas from young children.

The epitopic specificity of the natural response to Gtf and GbpB has been recently explored in groups of children during colonization of primary (5–13 months old) or permanent teeth (10–12 years old) with mutans streptococci (unpublished observations). Salivary IgA antibody levels were measured to 14 linear peptides derived from regions of suspected functional activity or MHC class II binding potential of Gtf and GbpB. Salivary antibody from either group of children reacted most consistently with peptides associated with the glucan-binding domain of Gtf. In addition, a salivary IgA-responsive epitope was often detected at an activity-associated site in the central portion of the enzyme. Salivary IgA reaction with the glucan-binding domain is also observed as a consequence of experimental infection of rats with mutans streptococci. Epitopes in the N terminal third of GbpB were most prominent in the salivary IgA immune response to this protein in both groups of children. This pattern was also reflected in the rodent response to infection with *S. mutans*. These observations not only provide more natural immune response detail, but also permit informed targeting in the design of subunit vaccines for dental caries, or in preparation of monoclonal passive immune reagents.

Permanent colonization with mutans streptococci leading to accumulation apparently is not always required for the induction of detectable levels of salivary IgA antibody to species-specific antigens. Antibody reactivity with mutans streptococcal antigens can be observed independent of the ability to detect ongoing infection in some children (Smith et al. 1998; Nogueira et al. 2005). As is the case with many bacterial challenges throughout the body, the threshold of immunological response is lower than that of persistent infection; therefore it is not surprising to observe antibody to *S. mutans* antigens in the absence of its colonization, since most children are being exposed to these organisms by their primary caregivers from birth. Thus colonization, at least not extensive colonization with mutans streptococci is not necessary for the development of salivary antibody to associated mutans streptococcal antigens. As indicated above, the relatively late appearance of these *S. mutans* antibody specificities suggests that they may not be primarily the result of cross-reactive responses to earlier colonizing microbiota.

Salivary IgA antibody responses to mutans streptococci show significant individual characteristics in early childhood. Children respond at different rates following infection, a condition which may be partly the result of the extent of infection (antigen dose) or the age at the time of infection (maturation of immune response). Siblings may also differ in their responses to MS antigens, despite being initially challenged with the same maternal subspecies. For example, Fig. 3 shows Western blots of the IgA reactivity patterns in two young siblings' (S1 and S2) salivas with culture supernatant antigen from the *S. mutans* isolated from S2. The patterns are as different as those seen in two unrelated adults (A1 and A2). Children may be challenged with several mutans streptococcal strains (Saarela et al. 1993), primarily by vertical (Li and Caufield 1995; Kozai et al. 1999), but also by horizontal (Mattos-Graner et al. 2001b) transmission. However, they become colonized with only some, one, or none, of these strains, a process which is influenced by salivary protein characteristics, epithelial receptors, and preexisting microbiota. The rate, specificity and/or extent of the mucosal immune response to previous encounters with the organism may also contribute to the success or failure of permanent colonization.

Nogueira and co-workers (2005) have evidence suggesting that vigorous immune responses to antigens important to colonization/accumulation may delay *S. mutans* infection under certain circumstances. Five- to 13-month-old children were paired with respect to age, number of teeth, IgA concentration and racial background, but differed in the presence or absence of *S. mutans*. All children were

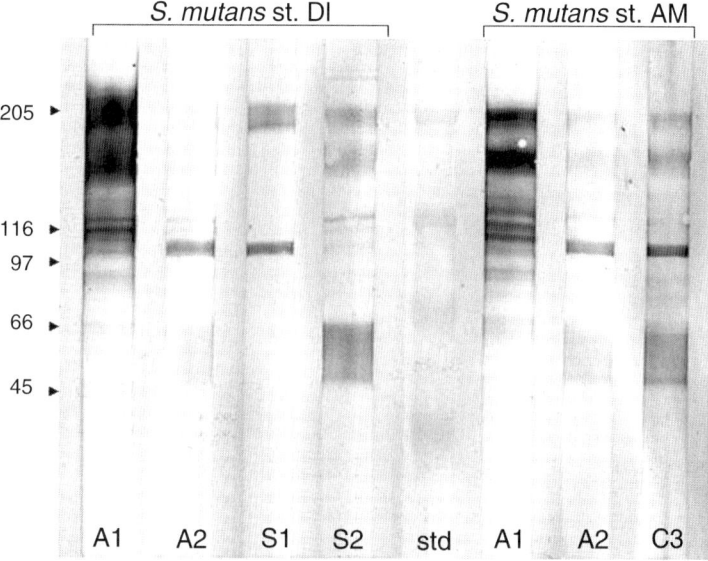

Fig. 3 Western blot of reactions of culture supernatants of two different *Streptococcus mutans* isolates (strains DI and AM) with IgA antibody in salivas from one sibling pair (SI and S2) who were 23 and 36 months of age, respectively, at the time of collection. Salivary IgA antibody reactivity with *S. mutans* antigens is also shown for a third, unrelated child (C3, 30 months of age) and two adults (A1 and A2). Migration of standards is indicated on the left edge of the blot by *arrows* and the corresponding kDa

considered to be under heavy exposure to oral mutans streptococcal challenge, based in the high levels of infection of primary caregivers and sucrose-rich diets (Nogueira et al. 2005). Antibody reactivity with Gtf and antigen I/II was detected in most of the children regardless of the infection status. However, salivary IgA antibody reactive with GbpB was observed in only 38% of infected children, while 76% of the uninfected children had Western blot IgA antibody reactivity with GbpB. Furthermore, among the anti-GbpB-positive children, the amount of densitometrically determined IgA antibody was generally much higher in the uninfected group, suggesting that GbpB was immunodominant in young children before detectable levels of mutans streptococci could be established in the oral cavity. Thus, children under heavy challenge with mutans streptococcal antigens can mount secretory IgA responses to *S. mutans*-specific antigens in the 1st year of life. In addition, these data suggest that the specificity of this natural response has the potential to influence colonization. Exploring factors which may define patterns of natural response to GbpB may provide important insights for the application of this antigen in anti-caries vaccines.

Taken together, the evidence from salivary IgA responses to commensal and pathogenic oral microorganisms indicates that the mucosal immune system is well developed by the period during which children typically become infected with mutans streptococci. Most children apparently respond immunologically to transient infection or ongoing colonization with mutans streptococci in early childhood. Although the distribution and specificity of children's responses are not identical, antibody to a few major antigens predominates. The possibility exists that such responses could be protective if induced prior to critical colonization events.

7 Active Immunization

Several strategies have been employed to test the hypothesis that the presence of antibody prior to infection can interfere with infection and disease. Virtually all approaches target either *Streptococcus mutans* or *S. sobrinus* as the principal etiologic agents of the disease. Most investigators sought to prevent initial colonization rather than to remove cariogenic microorganisms from an established biofilm. Bacterial colonization or accumulation was targeted by vaccines for inactivation/blockage, rather than metabolic pathways critical to the survival of the microorganism.

7.1 Targeting Adherence

Following early demonstrations that immunization with intact mutans streptococci could induce protective immune responses in experimental models for dental caries (reviewed in Michalek and Childers 1990), several investigators began to focus on antigens involved in adherence of *S. mutans* since the first step in the molecular pathogenesis of the disease is the colonization of the tooth surface, or, more accurately, colonization of the pre-existing biofilm on the tooth surface, through adhesin-mediated mechanisms.

Abundant in vitro and in vivo evidence indicates that antibody with specificity for *S. mutans* antigen I/II (AgI/II) or *S. sobrinus* SpaA can interfere with bacterial adherence and subsequent dental caries. Antibody directed to the intact Ag I/II molecule or to its salivary-binding domain blocked adherence of *S. mutans* to saliva-coated hydroxyapatite (reviewed in Koga et al. 2002). Furthermore, a variety of immunization approaches have shown that active immunization with intact Ag I/II, or passive immunization with monoclonal or transgenic antibody (see Sect. 9) to putative salivary binding domain epitopes within this component can protect rodents, primates, or humans from infection or disease caused by *S. mutans*. Immunization of mice with synthetic peptides (residues 301–319) from the alanine-rich region of Ag I/II suppressed tooth colonization with *S. mutans* (Takahashi et al. 1990). Intranasal immunization with Ag I/II, coupled to cholera toxin B subunit, reduced dental caries by *S. mutans* in an experimental rat model (Hajishengallis et al. 1998). Fusion proteins containing this adhesin were also shown to inhibit sucrose-independent adhesion of *S. mutans* to saliva-coated hydroxyapatite beads (Yu et al. 1997). Likewise, intranasally applied DNA vaccines encoding Ag I/II sequence were shown to be protective (Jia et al. 2004). Immunization with *S. sobrinus* SpaA constructs protected rats from caries caused by *S. sobrinus* infection (Redman et al. 1996). Thus, these streptococcal adhesins appear to be candidates for dental caries vaccine applications.

7.2 *Targeting Accumulation*

Sucrose-dependent accumulation of mutans streptococci within the dental biofilm has also been immunologically targeted in preclinical and clinical studies. Most effort has centered on a group of glucosyltransferase enzymes which catalyze the extracellular formation of alpha 1,3 and alpha 1,6-linked glucan from sucrose. These polysaccharides provide multiple binding sites for the accumulation of mutans streptococci in the plaque as well as serving as reservoirs of carbohydrate for cell metabolism. Given the central role of glucan formation in the molecular pathogenesis of dental caries, it is not surprising that Gtfs from *S. mutans* or *S. sobrinus* could induce protective immune responses in preclinical (reviewed in Smith 2002) or clinical studies in young adults (Smith and Taubman 1987). Thus, the presence of antibody to glucosyltransferase in the oral cavity prior to infection can significantly influence the disease outcome, presumably by interference with one or more of the functional activities of the enzyme.

Immune responses to Gtf can also be directed to functional domains. A variety of biochemical, genetic, and sequence alignment techniques have identified several residues within the N terminal region of Gtfs that appear to be associated with its catalytic activity. Other studies have shown that the repeating sequences in the C terminal third of Gtf are associated with primer-dependent binding of glucan (reviewed in Smith 2002). This information has directed the design of Gtf subunit vaccines. For example, immunization with synthetic peptide constructs or recombinant peptides corresponding to sequences containing at least five different catalytically active residues or glucan-binding domains of Gtf have been shown to induce immune responses interfering with enzyme function and/or with the cariogenic activity of mutans streptococcal infection (Jespersgaard et al. 1999; Smith 2002). Protection has also been observed recently with Gtf peptides associated with MHC class II-binding (Culshaw et al. 2007).

Glucan-binding proteins have also received attention in the search for effective antigens. Although several glucan-binding proteins (Gbp) have been described, only one, *S. mutans* GbpB, has been reported to induce protective immunity in experimental systems (Smith and Taubman 1996). Bioinformatic analysis of the GbpB sequence has revealed several sequences with potential MHC class II-binding activity. A synthetic peptide based on one of these sequences in the N terminal third of the molecule has been shown to induce protective immunity in a rat model for dental caries (Smith et al. 2002).

The ability to induce protective immunity with subunit vaccines based on adhesin, Gtf, or GbpB sequences has led investigators to develop constructs combining epitopes from one or more proteins. Taubman and co-workers (2001) have shown that the combination of peptides from the catalytic and glucan-binding domains of Gtf enhances the level of experimental enzyme and caries inhibition. Fusion proteins constructed from the saliva-binding alanine-rich region of Ag I/II and the glucan-binding domain of Gtf induced immune responses that inhibited water-insoluble glucan synthesis by *S. mutans* Gtf and also inhibited sucrose independent adhesion of *S. mutans* to saliva-coated hydroxyapatite beads (Yu et al. 1997). DNA vaccines coding for both the adhesin and Gtf also have been shown to induce protective responses (Guo et al. 2004). Incorporation of a Gtf catalytic domain peptide within a construct containing a GbpB sequence with MHC class II-binding characteristics greatly enhanced antibody formation to the Gtf peptide (Smith et al. 2005). Thus the potential exists to broaden the protective potential of the subunit vaccine approach.

8 Methods of Antigen Delivery and Immune Enhancement

8.1 Routes

Although the basic principle of immune protection from dental caries caused by mutans streptococci has been established in preclinical studies, refinements to the effective application of this approach to humans remain. One of the issues deals with the route of delivery. Early studies with mucosally applied caries vaccines used the oral or intragastric route for antigen delivery. Although significant effects on mutans streptococcal infection and disease were observed in animal and human studies (reviewed in Smith et al. 2002), induction of immune responses by the oral route requires antigen passage through the gut prior to uptake in the gut-associated lymphoid tissue (GALT). Since transit through the acidic stomach environment reduces the effective antigen stimulus, and the journey to GALT dilutes antigen concentration, induction of protective immunity in mucosal sites that are in closer anatomical relationship to the oral cavity has been pursued. Intranasal installation (IN) of antigen, which targets the nasal-associated lymphoid tissue (NALT), has been used to induce experimental immunity to many bacterial antigens, including those associated with mutans streptococcal colonization and accumulation. Protective immunity after infection with cariogenic mutans streptococci could be induced in rats by the IN route with many *S. mutans* antigens or functional domains associated with these components. Protection via this route could be demonstrated with *S. mutans* Ag I/II, the salivary binding region (SBR) of Ag I/II, a 19-mer sequence within the SBR, the glucan-binding domain of *S. mutans* GtfB, *S. mutans* GbpB or GbpB-derived peptides (reviewed in Koga et al. 2002; Smith 2002), and fimbrial preparations from *S. mutans* (Fontana et al. 1999), using antigen alone or in combination with mucosal adjuvants. Additional routes have included exposure of antigen to tonsils (Childers et al. 2002), to minor salivary glands (Smith and Taubman 1990), or rectal tissue (Smith et al. 2003b).

8.2 Adjuvants

Mucosal application of soluble antigen generally is insufficient to sustain an immune response. Thus, several enhancement strategies have been employed. One of these has been to take advantage of the immunostimulating ability of enterotoxins such as cholera toxin (CT) or the closely related serogroups I and II of *E. coli* heat labile enterotoxins (LT). CT could be shown to markedly enhance the immune response and/or the protective potential of intragastrically or intranasally applied mutans streptococcal antigens (Russell et al. 1991; Martin et al. 2000; Smith et al. 2001a). Adjuvant features of CT or LT for use in preclinical studies of dental caries vaccines were preserved by removing the A1 subunit toxin domain from the CT complex, mutating residues in the A subunit domain, or using only the nontoxic B subunit. Antigen or functional domains therein, when mixed with detoxified adjuvant or chemically conjugated or genetically fused with adjuvant subunits, dramatically increased salivary immune responses, resulting in protective immunity (reviewed in Smith 2002). However, refinements of these approaches are required because of potential retrograde transfer of these GM1-binding molecules to neuronal tissues following intranasal application (van Ginkel et al. 2000).

8.3 Delivery Systems

Targeting vaccine antigens to mucosal immune tissue or delivery in vehicles which prolong or promote exposure of antigen to inductive tissue has also been used successfully in preclinical studies. For example, intact adhesins, chimeric SBR-CTA2/B proteins (Hajishengallis et al. 1996), or Gtf glucan-binding domains (Taubman et al. 2004) were expressed in *Salmonella*, since these bacteria specifically bind to the epithelial cells which overlie mucosal lymphoid tissue. Such constructs, delivered in attenuated *Salmonella* expression vectors, induced protective immune responses when administered intragastrically or intranasally. Incorporation of antigen in or on various types of microparticles or microspheres made of poly(lactide-co-glycolide) or liposomes has also been employed in an attempt to increase uptake via antigen particularization. These approaches have been shown to be useful in the delivery of dental caries vaccine antigens for protective responses in preclinical studies (reviewed in Smith 2002) or salivary antibody formation in humans (Childers et al. 1999).

9 Passive Immunization

As an alternative to strategies that require active immune responses in the host, antibody to critical *S. mutans* virulence factors can be passively introduced into the oral cavity for protective effect (reviewed in Koga et al. 2002). Two pioneering studies established this concept. The first study, in which monkeys were infused intravenously with simian IgG antibody directed to *S. mutans* cell surface antigens, indicated that non-host IgG antibody entering the oral cavity via the gingival crevicular fluid, could reduce dental caries in rhesus monkeys (Lehner et al. 1978). A second pioneering study showed that mucosally derived antibody could also impart protection. In this investigation, Michalek and McGhee (1977) immunized rat dams with mutans streptococcal cells, resulting in mucosal secretion of antibody to *S. mutans* in their milk. Offspring of these dams, after suckling on antibody-rich milk, showed a significant degree of protection from dental caries caused by subsequent infection with *S. mutans*. These experiments established the concept of passive protection for dental caries and again showed that the specificity and presence of antibody in the oral cavity was more important for protection than the method by which antibody entered the oral cavity.

Some (Otake et al. 1991), but not all (Hamada et al. 1991) subsequent investigations, in which rats were given continuous dietary exposure to antibody to whole *S. mutans* cells, supported the passive approach. Modest, short-term reductions in plaque-associated *S. mutans* were also observed when bovine antibody to whole cells was incorporated into the diet of gnotobiotic rats (Michalek et al. 1987) or used as a mouth rinse in humans (Loimaranta et al. 1999).

Improved protection followed the targeting of the antibody reagents to individual components of the mutans streptococcal colonization or accumulation processes. For example, polyclonal IgG antibody or IgG monoclonal reagents to surface adhesins of *S. mutans* (Lehner et al. 1985) or *Streptococcus sobrinus* (van Raamsdonk et al. 1993) could be shown to reduce colonization and inhibit dental caries in Rhesus monkeys, or reduce colonization in rats, when administered by repeated topical application. Human application of the passive immunization approach was highlighted in several small clinical trials with monoclonal reagents directed against an epitope of *S. mutans* antigen I/II. In these experiments, teeth in adult volunteers were treated for 9 days with chlorhexidine to remove most of the dental microbiota. This antibacterial treatment was followed by a 3-week regimen in which mouse monoclonal IgG (Ma et al. 1990) or transgenic SIgA/G (Ma et al. 1996) antibody to an adhesin epitope, prepared in tobacco plants, was topically applied to the teeth. Recolonization of the tooth surfaces with *S. mutans* did not occur in adhesin antibody-treated subjects over the course of the study, in contrast to the tooth surfaces of subjects treated with antibody irrelevant to *S. mutans*. The working hypothesis for these observations was that non-mutans streptococcal members of the oral biofilm were given a selective advantage in filling the ecological niche previously occupied by *S. mutans* as a result of passive antibody treatment, thus blocking mutans streptococcal recolonization. A subsequent clinical trial (Weintraub et al. 2005) employing a somewhat different protocol but using a similar transgenic antibody reagent, replicated the safety, but not the effectiveness reported by Ma and co-workers (1996).

A novel method for delivery of relevant monoclonal antibody specificities in the oral cavity has been reported by Kruger and co-workers (2002, 2005). In this approach, vectors encoding a single-chain Fv (scFc) fragment of antibody with specificity for the Ag I/II adhesin epitope were expressed in *Lactobacillus zeae* (formerly, *casei*) so that the scFv was displayed on the lactobacillus surface. When young, desalivated rats were orally swabbed with these transformed lactobacilli throughout the course of *S. mutans* infection, significant reductions in infection and dental caries were observed. The authors suggest that the surface display of svFc on lactobacilli results in a multivalent entity, which would be predicted to have the avidity and valence to prevent (re)colonization of the teeth by *S. mutans*.

Very recent efforts have explored the ability to enhance the protective effect of passively applied antibody by fusing it to a bacteriocidal component, in this case, glucose oxidase (Kruger et al. 2006). Glucose oxidase enhances the peroxidase system through the generation of hydrogen peroxide, which can then be converted to hypothiocyanite and hypoiodite, both of which have microbiocidal properties. In these studies, antibody specificity to *S. mutans* was generated in the variable domain of llama heavy-chain homodimeric IgG (VHH). Fusion proteins containing the VHH and glucose oxidase sequences are expressed in *Streptocmyces cerevisiae*. These fusion proteins are then applied daily to the oral cavities of desalivated rats that had been infected with *S. mutans*. Although reductions in dental caries were modest, perhaps because of the lack of peroxidase-containing saliva or less than optimal antibody specificity, this general approach may have useful future application.

Experimental approaches also demonstrated the value of passive antibody administration to other primary immune targets in the molecular pathogenesis of dental caries. Most used antibody derived from chicken egg yolk as the passive antibody source. The yolk of the chicken egg contains high concentrations of an IgG-like protein referred to as IgY. Antibody of the IgY immunoglobulin class reflects the immunological experience of the hen. Thus, given regular egg production, immunization of a limited number of birds can result in recovery of significant amounts of IgY antibody to the immunizing agent. This production capacity makes the IgY approach attractive for the quantities of antibody required for the passive immunization approach. In addition, approval for human use may be facilitated since egg products are already a common food source.

Hamada and co-workers (1991) prepared IgY antibody to cell-associated Gtf (CA-Gtf) or Gtf from the culture supernatants (CF-Gtf) of *S. mutans* MT8148R. In vitro analyses revealed that the more purified IgY antibody to CA-Gtf significantly inhibited glucan formation by CA-Gtf and could interfere with sucrose-dependent adherence of *S. mutans*. These features then served as the theoretical bases for the protective effects seen in vivo. Significant reductions in plaque deposition and caries development occurred in the rats fed purified IgY antibody to CA-Gtf at concentrations up to 1% of the diet. Interestingly, dietary administration of antibody to Cf-Gtf did not result in demonstrable protection, presumably because this antibody reagent was significantly less inhibitory of Gtf-mediated glucan synthesis than was antibody to CA-Gtf, and did not inhibit in vitro sucrose-dependent adherence of *S. mutans*. More recently, immune milk containing antibody to a fusion protein consisting of the alanine-rich salivary binding region of antigen I/II and the glucan binding domain of Gtf was used as a mouth rinse (Shimazaki et al. 2001). Following cetylpyridinium treatment of the teeth of adults, and mouth rinsing with immune milk, significant reductions in recolonizing mutans streptococci were reported.

This IgY approach was extended by Kruger and co-workers (2004), who investigated the influence of IgY antibody to CA-Gtf in rats who were surgically desalivated. This application is of interest because salivary flow has been directly associated with oral health, including dental caries. Those individuals whose salivary flow has been compromised by diseases such as Sjögren's syndrome or by head and neck irradiation are, therefore, at increased risk for dental caries. The group of rats drinking water containing IgY CA-Gtf antibody had significantly lower caries scores and less severe lesions on both smooth and sulcal surfaces than did rats drinking non-immune IgY. The authors suggested that these data indicate that the IgY antibody not only prevented the initiation of caries, but also their progression. Thus, passive provision of antibody to at least some epitopes of *S. mutans* GTF in the diet or drinking water can provide effective protection under experimental conditions in the rat.

Antibody to *S. mutans* glucan-binding protein B (GbpB) was also shown to provide passive protection against experimental dental caries (Smith and Godiska 2001), even when antibody exposure occurred only in the initial stage of infection. In these experiments, IgY antibody to GbpB, or pre-immune IgY, was added to cariogenic diet 2000 for 9 or 23 days and given to experimental or control groups, respectively. Periodic measurement of the extent of infection with *S. mutans* indicated that IgY antibody to GbpB had a significant effect on accumulation of these cariogenic streptococci throughout the 78-day infection period. At the end of this infection period, rats fed for 9 days had significant dental caries reductions on smooth molar surfaces, while in rats fed for 23 days, reductions were significant on both occlusal and smooth surfaces. Furthermore, when caries scores were calculated with respect to individual teeth, significant reductions were observed on eight of 12 teeth in the second experiment. Interestingly, the level of caries protection after 23 days of IgY antibody feeding was similar to that seen in experiments where rats were actively immunized with GbpB followed by a similar *S. mutans* infection period (Smith et al. 1997). Thus, early exposure to dietary IgY antibody to *S. mutans* GbpB can diminish the extent of eventual dental disease, and possibly bacterial accumulation, prior to observable disease. Furthermore, relatively short-term administration of IgY antibody to GbpB was shown to be essentially as effective as active immunization in inhibiting caries formation in this experimental system.

10 Summary

Salivary IgA antibody responses to mutans streptococci can be observed in early childhood, sometimes even before permanent colonization of the oral biofilm occurs. Many of these early immune responses are directed to components thought to be essential for establishment and emergence of mutans streptococci in the oral biofilm. Initial responses are likely to be modulated by antigen dose, by immunological maturity and by previous encounters with similar antigenic epitopes in the pioneer commensal flora. Our understanding of these modulating factors is modest and is an opportunity for continued investigation. Under controlled conditions of infection, experimental vaccine approaches have repeatedly shown that infection and disease can be modified in the presence of elevated levels of antibody in the oral cavity. Protection can be observed regardless of antibody isotype or method used to actively or passively provide the immune reagent. Limited clinical trials have supported the utility of both of these approaches in humans. Refinements in antigen formulation, delivery vehicles, enhancing agents, and routes of application, coupled with approaches that are timed to intercept most vulnerable periods of infection of primary and permanent dentition may well provide the health care practitioner with an additional tool to maintain oral health.

Acknowledgements Grant support for the authors' research has come from the U.S. Public Health Service (DE-06133, TW-06324, DE-04733, DE/AI-12434), and Fundação de Amparo à Pesquisa do Estado de São Paulo (Brazil) (proc. 99/08278–9, 02/07156–1, 03, 03/01836–3), and CAPES (Brazil) (029/03).

References

Alaluusua S, Renkonen OV (1983) *Streptococcus mutans* establishment and dental caries experience in children from 2 to 4 years old. Scand J Dent Res 91:453–457

Banas JA, Vickerman MM (2003) Glucan binding proteins of the oral streptococci. Crit Rev Oral Biol Med 14:89–99

Banas JA, Russell RR, Ferretti JJ (1990) Sequence analysis of the gene for the glucan-binding protein of *Streptococcus mutans* Ingbritt. Infect Immun 58:667–673

Beisner DR, Ch'en IL, Kolla RV, Hoffmann A, Hedrick SM (2005) Cutting edge: innate immunity conferred by B cells is regulated by caspase-8. J Immunol 175:3469–3473

Brandtzaeg P (2007) Induction of secretory immunity and memory at mucosal surfaces. Vaccine 25:5467–5484

Brandtzaeg P, Johansen F-E (2005) Mucosal B cells: phenotypic characteristics, transcriptional regulation, and homing properties. Immunol Rev. 206:32–63

Burne RA (1998) Oral streptococci... products of their environment. J Dent Res 77:445–452

Camling E, Gahnberg L, Krasse B, Wallman C (1991) Crevicular IgG antibodies and *Streptococcus mutans* on erupting human first permanent molars. Arch Oral Biol 36:703–708

Camling E, Gahnberg L, Emilson CG, Lindquist B (1992) Crevicular IgG antibodies and recovery of *Streptococcus mutans* implanted by mouth rinsing. Scand J Dent Res 6:134–138

Catalanotto FA, Shklair IL, Keene HJ (1975) Prevalence and localization of *Streptococcus mutans* in infants and children. J Am Dent Assoc 91:606–609

Caufield PW, Cutter GR, Dasanayake AP (1993) Initial acquisition of mutans streptococci by infants: evidence for a discrete window of infectivity. J Dent Res 72:37–45

Challacombe SJ, Percival RS, Marsh PD (1995) Age-related changes in immunoglobulin isotypes in whole and parotid saliva and serum in healthy individuals. Oral Micro Immunol 10:202–207

Childers NK, Tong G, Mitchell S, Kirk K, Russell MW, Michalek SM (1999) A controlled clinical study of the effect of nasal immunization with a *Streptococcus mutans* antigen alone or incorporated into liposomes on induction of immune responses. Infect Immun 67:618–623

Childers NK, Tong G, Li F, Dasanayake AP, Kirk K, Michalek SM (2002) Humans immunized with *Streptococcus mutans* antigens by mucosal routes. J Dent Res 81:48–52

Cole MF, Bryan S, Evans MK, Pearce CL, Sheridan MJ, Sura PA, Wientzen RL Bowden GHW (1999) Humoral immunity to commensal oral bacteria in human infants: salivary secretory immunoglobulin A antibodies reactive with *Streptococcus mitis* biovar 1, *Streptococcus oralis*, *Streptococcus mutans*, and *Enterococcus faecalis* during the first two years of life. Infect Immun 67:1878–1886

Culshaw S, LaRosa K, Tolani H, Han X, Eastcott JW, Smith DJ, Taubman MA (2007) Immunogenic and protective potential of mutans streptococcal glucosyltransferase peptide constructs selected by major histocompatibility complex class II allele binding. Infect Immun 75:915–923

Fitzsimmons SP, Evans MK, Pearce CL, Sheridan MJ, Wientzen R, Cole MF (1994) Immunoglobulin A subclasses in infants' saliva and in saliva and milk from their mothers. J Pediatr 124:566–573

Fontana M, Dunipace AJ, Stookey GK, Gregory RL (1999) Intranasal immunization against dental caries with a *Streptococcus mutans*-enriched fimbrial preparation. Clin Diagn Lab Immunol 6:405–409

Guo JH, Jia R, Fan MW, Bian Z, Chen Z, Peng B (2004) Construction and immunogenic characterization of a fusion anti-caries DNA vaccine against PAc and glucosyltransferase I of *Streptococcus mutans*. J Dent Res 83:266–270

Hajishengallis G, Harokopakis E, Hollingshead SK, Russell MW, Michalek SM (1996) Construction and oral immunogenicity of a *Salmonella typhimurium* strain expressing a streptococcal adhesin linked to the A2/B subunits of cholera toxin. Vaccine 14:1545–1548

Hajishengallis G, Russell MW, Michalek SM (1998) Comparison of an adherence domain and a structural region of *Streptococcus mutans* antigen I/II in protective immunity against dental caries in rats after intranasal immunization. Infect Immun 66:1740–1743

Hamada S, Slade HD (1980) Biology, immunology and cariogenicity of *Streptococcus mutans*. Microbiol Rev 44:331–384

Hamada S, Horikoshi, T, Minami T, Kawabata S, Hiraoka J, Fujiwara T, Ooshima T (1991) Oral passive immunization against dental caries in rats by use of hen egg yolk antibodies specific for cell-associated glucosyltransferase of *Streptococcus mutans*. Infect Immunity 59:4161–4167

Hatta H, Tsuda K, Ozeki M, Kim M, Yamamoto T, Otake S, Hirasawa M, Katz J, Childers NK, Michalek SM (1997) Passive immunization against dental plaque formation in humans: effect of a mouth rinse containing egg yolk antibodies (IgY) specific to *Streptococcus mutans*. Caries Res 31:268–274

Hayes JA, Adamson-Macedo EN, Perera S, Anderson J (1999) Detection of secretory immunoglobulin A (SIgA) in saliva of ventilated and non-ventilated preterm neonates. Neuro Endocrinol Lett 20:109–113

Jayashree S, Bhan MK, Kumar R, Raj P, Glass R, Bhandari N (1988) Serum and salivary antibodies as indicators of rotavirus infection in neonates. J Infect Dis 158:1117–1120

Jenkinson HF, Lamont R (2005) Oral microbial communities in sickness and in health. Trends Microbiol 13:589–595

Jespersgaard C, Hajishengallis G, Huang Y, Russell MW, Smith DJ, Michalek SM (1999) Protective immunity against *Streptococcus mutans* infection in mice after intranasal immunization with the glucan-binding region of *S. mutans* glucosyltransferase. Infect Immunity 67:6543–6549

Kent R, Smith DJ, Joshipura K, Soparkar P, Taubman MA (1992) Humoral IgG antibodies to oral microbiota in a population at risk for root-surface caries. J Dent Res 71:1399–1407

Klein MI, Florio FM, Pereira AC, Hofling JF, Goncalves RB (2004) Longitudinal study of transmission, diversity, and stability of *Streptococcus mutans* and *Streptococcus sobrinus* genotypes in Brazilian nursery children. J Clin Microbiol 42:4620–4626

Koga T, Oho T, Shimazaki Y, Nakano Y (2002) Immunization against dental caries. Vaccine 20:2027–2044

Kohler B, Andreen I, Jonsson B (1988) The earlier the colonization by mutans streptococci, the higher the caries prevalence at 4 years of age. Oral Microbiol Immunol 3:14–17

Kruger C, Hu Y, Pan Q, Marcotte H, Hultberg A, Delwar D, van Dalen PJ, Puwels PH, Leer RJ, Kelly CG, van Dollenweerd C, Ma JK, Hammarstrom L (2002) *In situ* delivery of passive immunity by lactobacilli producing single-chain antibodies. Nature Biotech 20:702–706

Kruger C, Pearson SK, Kodama Y, Vacca Smith A, Bowen WH, Hammarstrom L (2004) The effects of egg-derived antibodies to glucosyltransferases on dental caries in rats. Caries Res 38:9–14.

Kruger C, Hultberg A, van Dollenweerd C, Marcotte H, Hammarstrom L (2005) Passive immunization by lactobacilli expressing single-chain antibodies against *Streptococcus mutans*. Mol Biotechnol 31:221–231

Kruger C, Hultberg A, Marcotte H, Hermans P, Bezemer S, Frenken LGJ, Hammarstrom L (2006) Therapeutic effect of llama derived VHH fragments against *Streptococcus mutans* on the development of dental caries. Appl Microbiol Biotechnol 72: 732–737

Jia R, Gao JH, Fan MW, Bian Z, Chen Z, Peng B, Fan B (2004) Mucosal immunization against dental caries with plasmid DNA encoding *pac* gene of *Streptococcus mutans* in rats. Vaccine 22:2511–2516

Johansen F-E, Baekkevold ES, Carlsen HS, Farstad IN, Soler D, Brandtzaeg P (2005) Regional induction of adhesion molecules and chemokine receptors explains disparate homing of human B cells to systemic and mucosal effector sites: dispersion from tonsils. Blood 106:593–600

Lehner T, Russell MW, Challacombe SJ, Scully CM, Hawkes JE (1978) Passive immunization with serum and immunoglobulins against dental caries in rhesus monkeys. Lancet 1:693–695

Lehner T, Caldwell J, Smith R (1985) Local passive immunization by monoclonal antibodies against streptococcal antigen I/II in the prevention of dental caries. Infect Immunity 50:796–799

Li Y, Caufield PW (1995) The fidelity of initial acquisition of mutans streptococci by infants from their mothers. J Dent Res 74:681–685

Loesche WJ (1986) Role of *Streptococcus mutans* in human dental decay. Microbiol Rev 50:353–380

Loimaranta V, Laine M, Soderling E, Vasara E, Rokka S, Marnila P, Korhonen H, Tossavainen O, Tenovuo J (1999) Effects of bovine immune and non/immune whey preparations on the composition and pH responses of human dental plaque. Eur J Oral Sci 107:244–250

Luo Z, Smith DJ, Taubman MA, King WF (1988) Cross sectional analysis of serum antibody to oral streptococcal antigens in children. J Dent Res 67:554–560

Ma JK, Lehner T (1990) Prevention of colonization of *Streptococcus mutans* by topical application of monoclonal antibodies in human subjects. Archs Oral Biol 35 Suppl:115S–122S

Ma JK, Hunjan M, Smith R, Kelly C, Lehner T (1990) An investigation into the mechanism of protection by local passive immunization with monoclonal antibodies against *Streptococcus mutans*. Infect Immunity 58:3407–3414

Ma JK, Hikmat BY, Wycoff K, Vine ND, Chargelegue D, Yu L, Hein MB, Lehner T (1998) Characterization of a recombinant plant monoclonal secretory antibody and preventive immunotherapy in humans. Nature Med 4:601–606

Macpherson AJ, Gatto D, Sainsbury E, Harriman GR, Hengartner H, Zinkernagel RM (2000) A primitive T cell-independent mechanism of intestinal mucosal IgA responses to commensal bacteria. Science 288:2222–2226

Macpherson AJ, Geuking MB, McCoy KD (2005) Immune responses that adapt the intestinal mucosa to commensal intestinal bacteria. Immunology 115:153–162

Martin M, Metzger DJ, Michalek SM, Connell TD, Russell MW (2000) Comparative analysis of the mucosal adjuvanticity of the type II heat-labile enterotoxins LT-IIa and LT-IIb. Infect Immun 68:281–287

Mattos-Graner RO, Rontani RM, Gaviao MB, Bocatto HA (1996) Caries prevalence in 6–36-month-old Brazilian children. Community Dent Health 13:96–98

Mattos-Graner RO, Smith DJ, King WF, Mayer MPA (2000) Water-insoluble glucan synthesis by mutans streptococcal strains correlates with caries incidence in 12- to 30-month-old children. J Dent Res 79:1371–1377

Mattos-Graner RO, Correa MS, Latorre MR, Peres RC, Mayer MP (2001a) Mutans streptococci oral colonization in 12–30-month-old Brazilian children over a one-year follow-up period. J Public Health Dent 61:161–167

Mattos-Graner RO, Li Y, Caufield PW, Duncan M, Smith DJ (2001b) Genotypic diversity of mutans streptococci in Brazilian nursery children suggests horizontal transmission. J Clin Microbiol 39:2313–2316

Mellander L, Carlsson B, Hanson LA (1984) Appearance of secretory IgM and IgA antibodies to *E. coli* in early infancy and childhood. J Pediatr 104:564–568

Michalek SM, Childers NK (1990) Development and outlook for a caries vaccine. Crit Rev Biol Med 1:37–54

Michalek SM, McGhee JR (1977) Effective immunity to dental caries: passive transfer to rats of antibody to *Streptococcus mutans* elicit protection. Infect Immunity 17:644–650

Newburg DS, Walker WA (2007) Protection of the neonate by the innate immune system of developing gut and of human milk. Pediatr Res 61:2–8

Nogueira RD, Alves AC, Napimoga MH, Smith DJ, Mattos-Graner RO (2005) Characterization of salivary IgA responses in children heavily exposed to the oral bacteria *Streptococcus mutans*: influence of specific antigen recognition in infection. Infect Immun 73:5675–5684

Otake S, Nishahara Y, Makinura M, Hatta H, Kim M, Yamamoto T, Hirasawa T (1991) Protection of rats against dental caries by passive immunization with hen-egg yolk antibody (IgY). J Dent Res 70:162–166

Percival RS, Marsh PD, Challacombe SJ (1997) Age-related changes in salivary antibodies to commensal oral and gut biota. Oral Micro Immunol 12:57–63

Quivey RG, Kuhnert WL, Hahn K (2001) Genetics of acid adaptation in oral streptococci. Crit Rev Oral Biol Med 12:301–314

Quan CP, Berneman A, Pires R, Avrameas S, Bouvet JP (1997) Natural polyreactive secretory immunoglobulin A autoantibodies as a possible barrier to infection in humans. Infect Immun 65:3997–4004

Redman TK, Harmon CC, Michalek SM (1996) Oral immunization with recombinant *Salmonella typhimurium* expressing surface protein antigen A (SpaA) of *Streptococcus sobrinus*: effects of the Salmonella virulence plasmid on the induction of protective and sustained humoral responses in rats. Vaccine 14:868–878

Russell MW, Lehner T (1978) Characterization of antigens extracted from cells and culture supernatant fluids of *Streptococcus mutans* serotype c. Archs Oral Biol 23:7–15

Russell MW, Wu HY (1991) Distribution, persistence, and recall of serum and salivary antibody responses to peroral immunization with protein antigen I/II of *Streptococcus mutans* coupled to the cholera toxin B subunit. Infect Immun 59:4061–4070

Russell MW, Hajishengallis G, Childers NK, Michalek SM (1999) Secretory immunity in defense against cariogenic mutans streptococci. Caries Res. 33:4–15

Saarela M, Alaluusua S, Takei T, Asikainen S (1993) Genetic diversity within isolates of mutans streptococci recognized by an rRNA gene probe. J Clin Microbiol 31:584–587

Sato Y, Yamamoto Y, Harutoshi K (1997) Cloning and sequence analysis of the GbpC gene encoding a novel glucan-binding protein of *Streptococcus mutans*. Infect Immun 65:668–675

Schour I, Massler M (1962) The Teeth. In: Farris E, Griffith J (eds) Hafner, New York, p 106

Seifert TB, Bleiweis AS, Brady LJ (2004) Contribution of the alanine-rich region of *Streptococcus mutans* P1 to antigenicity, surface expression, and interaction with the proline-rich repeat domain. Infect Immun 72:4699–4706

Senadheera MD, Guggenheim B, Spatafora GA, Huang YC, Choi J, Hung DC, Treglown JS, Goodman SD, Ellen RP, Cvitkovitch DG (2005) A VicRK signal transduction system in *Streptococcus mutans* affects *gtfBCD*, *gbpB*, and *ftf* expression, biofilm formation, and genetic competence development. J Bacteriol 187:4064–4076

Senadheera MD, Lee AW, Huang DC, Spatafora GA, Goodman SD, Cvitkovitch DG (2007) The *Streptococcus mutans* vicX gene product modulates gtfB/C expression, biofilm formation, genetic competence, and oxidative stress tolerance. J Bacteriol 189:1451–1458

Shah DS, Russell RR (2004) A novel glucan-binding protein with lipase activity from the oral pathogen *Streptococcus mutans*. Microbiology 150:1947–1956

Shimazaki Y, Mitoma M, Oho T, Nakano Y, Yamashita Y, Okano K, Fujiyama M, Fujihara N, Nada Y, Koga T (2001) Passive immunization with milk produced from an immunized cow prevents oral recolonization by *Streptococcus mutans*. Clin Diagn Lab Immunol 8:1136–1139

Smith DJ (2002) Dental caries vaccines: Prospects and concerns. Crit Rev Oral Biol Med 13:335–349

Smith DJ, Taubman MA (1987) Oral immunization of humans with *Streptococcus sobrinus* glucosyltransferase. Infect Immun 55:2562-2569

Smith DJ, Taubman MA (1987) Ontogeny and senescence of salivary immunity. J Dent Res 66:451–456

Smith DJ, Taubman MA (1990) Effect of local deposition of antigen on salivary immune responses and reaccumulation of mutans streptococci. J Clin Immunol 10:273-281

Smith DJ, Taubman MA (1992) Ontogeny of immunity to oral microbiota. Crit Rev Oral Biol Med 3:109–133

Smith DJ, Taubman MA (1993) Emergence of immune mechanisms in saliva. Crit Rev Oral Med 4:335–341

Smith DJ, Taubman MA (1996) Experimental immunization of rats with a *Streptococcus mutans* 59-kilodalton glucan-binding protein protects against dental caries. Infect Immun 64:3069–3073

Smith DJ, King WF, Taubman MA (1989) Isotype, subclass and molecular size of immunoglobulins in salivas from young infants. Clin Exper Immunol 76:97-102

Smith DJ, Akita H, King WF, Taubman MA (1994) Purification and antigenicity of a novel glucan binding protein of *S. mutans*. Infect Immun 62:2545–2552

Smith DJ, Heschel R, Melvin J, King WF, Pereira MBB, Taubman MA (1997) *Streptococcus mutans* glucan binding proteins as dental caries vaccines. In: Husband AJ, Beagley KW, Clancey RL, et al (eds) Mucosal Solutions: Advances in Mucosal Immunology. University of Sydney, Sydney, pp 367–377

Smith DJ, King WF, Akita H, Taubman MA (1998) Association of salivary IgA antibody and initial mutans streptococcal infection. Oral Micro Immunol 13:278–285

Smith DJ, King WF, Barnes LA, Trantolo D, Wise DL, Taubman MA (2001a) Facilitated intranasal induction of mucosal and systemic immunity to mutans streptococcal glucosyltransferase peptide vaccines. Infect Immun 69:4767–4773

Smith DJ, King WF, Godiska R (2001b) Passive transfer of immunoglobulin Y antibody to *Streptococcus mutans* glucan binding protein B can confer protection against experimental dental caries. Infect Immunity 69:3135–3142

Smith DJ, King WF, Barnes LA, Peacock Z, Taubman MA (2003a) Immunogenicity and protective immunity induced by synthetic peptides associated with putative immunodominant regions of *Streptococcus mutans* glucan-binding protein B. Infect Immun 71:1179–1184

Smith DJ, Lam A, Barnes LA, King WF, Peacock Z, Wise DL, Trantolo DJ, Taubman MA (2003b) Remote glucosyltransferase-microparticle vaccine delivery induces protective immunity in the oral cavity. Oral Micro Immunol 18:240–248

Smith DJ, King WF, Rivero J, Taubman MA (2005) Immunological and protective effect of diepitopic subunit dental caries vaccines. Infect Immun 73:2797–2804

Takahashi I, Okahashi N, Matsushita K, Tokuda M, Kanamoto T, Munekata E et al (1991) Immunogenicity and protective effect against oral colonization by *Streptococcus mutans* of synthetic peptides of a streptococcal surface protein antigen. J Immunol 146:332–336

Tanner AC, Milgrom PM, Kent R Jr, Mokeem SA, Page RC, Liao SI et al (2002) Similarity of the oral microbiota of pre-school children with that of their caregivers in a population-based study. Oral Microbiol Immunol 17:379–387

Tanzer JM (1995) Dental caries is a transmissible infectious disease: the Keyes and Fitzgerald revolution. J Dent Res 74:1536–1542

Tanzer JM, Livingston J, Thompson AM (2001) The microbiology of primary dental caries in humans. J Dent Educ 65:1028–1037

Taubman MA, Nash D (2006) The scientific and public health imperative for a dental caries vaccine. Nat Rev Immunol 6:555–563

Taubman MA, Holmberg CJ, Smith DJ (2001) Diepitopic construct of functionally and epitopically complementary peptides enhances immunogenicity, reactivity with glucosyltransferase, and protection from dental caries. Infect Immun 69:4210–4216

Taubman MA, Orr N, Smith DJ, Eastcott J, Hayden T (2002) A rodent model for analysis of immunization with GTF peptides expressed in attenuated *Salmonella enterica*. Mucosal Immunol Update. 10: 2553

van Ginkel FW, Jackson RJ, Yuki Y, McGhee JM (2000) Cutting edge: the mucosal adjuvant cholera toxin redirects vaccine proteins into olfactory tissues. J Immunol 165:4778–4782

Weintraub JA, Hilton JF, White JM, Hoover CI, Wycoff KL, Yu L, Larrick JW, Featherstone JDB (2005) Clinical trial of a plant-derived antibody on recolonization of mutans streptococci. Caries Res 39:241–250

Whiley RA, Beighton D (1998) Current classification of the oral streptococci. Oral Microbiol Immunol 13:195–216

Yu H, Nakano Y, Yamashita Y, Oho T, Koga T (1997) Effects of antibodies against cell surface protein antigen PAc-glucosyltransferase fusion proteins on glucan synthesis and cell adhesion of *Streptococcus mutans*. Infect Immun 65:2292–2298

Index

A
ActHIB, 26, 28
ACWY Vax, 26, 30
Adhesin, 135, 145, 146, 149
Adjuvants, 24, 35, 36
Altastaph, 27, 33, 34
Antibodies, 63, 64, 66–69, 71–88
Antibody effector mechanisms, 42
Antigen(s), 105–107, 113, 115–121, 123, 124
    density, 2–4, 7
    organization, 3–7
    T independent, 3
Antigenic competition, 21, 36
Anti-opsonic activity, 23

B
B cell antigen receptor (BCR), 106, 107, 110, 115–118, 120, 123
B cell memory, 105, 106, 121, 123, 124
B cell receptor (BCR), 19–21, 24
B cells subsets, 107–108, 110, 113, 114, 116, 120, 121, 123
    B-la cells, 43, 46–51
    B-lb cells, 43, 46–51
    follicular cells, 43–44
    Marginal zone (MZ) cells, 43–46
B-1 B cell, 23
B-1 cell
    regulation, 50
B1b cells, 105, 106, 108–124
B-2 cells, 43, 46, 47, 55, 56
Bacteremia, 109, 110, 112, 113, 117
Borrelia, 63–70, 72–87
Borrelia burgdorferi, 66, 68–72, 74–82, 84–86
Borrelia hermsii, 51, 56, 109–115, 117–119, 121, 123
Borrelia immunology, 75, 78

Bruton's tyrosine kinase (BTK), 106, 115–119
BSYX-A110, 27, 34

C
C3, 117, 118, 120, 121
Capsules, 113, 115
α/β CD4 + T cells, 24
CD19, 113, 120
CD40, 116, 117
Chemokine, 108, 109
Children, 120, 121
Class-switching, 9–10
Complement, 63, 66–69, 74, 75, 77, 83–88
Complement-independent bactericidal antibodies, 69, 84–88
COMVAX, 28
Conjugate vaccines, 21, 22, 25–28, 30, 32–36
Convalescent, 109, 110, 112, 121–123
CR1/2, 118, 120

D
Dental caries, 133, 134, 136–138, 142, 144–146, 148–151

G
Germinal center, 105–107, 110, 112, 115, 123, 124
Glucan binding protein, 135, 136, 146, 150
Glucosyltransferase, 135, 142, 145, 146, 148, 150

H
Haemophilus influenzae type b (Hib), 24–29
Hiberix, 26, 28
HibTITER, 26, 28

Host response, 63, 68–70, 73, 74, 77, 80, 84
Humoral immunity, 66, 68, 71
4-hydroxy-3-nitrophenyl acetyl (NP), 106, 114–116, 119, 120, 123

I

IgM, 69, 70, 72–74, 76, 77, 79–82, 85, 108–114, 116–124
Immunity, 66, 68–73, 75, 76, 80, 84–86, 106, 108, 109, 112, 113, 119, 121, 123, 124
Immunologic memory, 24, 28, 29
Influenza
    evasion, 51
    humoral response, 41, 51–52
Innate early B cell activation, 52

L

Lipid rafts, 19
Lipopolysaccharide (LPS), 106, 107, 116, 119
Lipoproteins, 20
Lipoteichoic acid (LTA), 27, 34–35
Long-lasting, 105, 107, 112–114, 119, 121, 123, 124
LPS stimulation, 49, 57
Lyme disease, 63, 64, 68–72, 74, 76–84
Lymph node, 108, 109, 114

M

Marginal zone (MZ) B cells, 23
    development, 44–45
    extrafollicular foci response, 45
    germinal center response, 45
Menactra, 26, 30
Meningitec, 26
Menjugate, 26, 30
Menomune-A/C/Y/W-135, 26, 30
MHC classII, 20, 22, 23
MyD88, 116, 117, 119, 120

N

Naïve, 108, 110, 112, 121, 123, 124
Natural antibody, 47, 55
Neisseria meningitides, 25, 29–31
NeisVac-C, 26, 30
Nonclassical T cells, 25
Nontypeable *H. influenzae*, 29, 32

O

Ontogeny, 138–141

P

Pagibaximab, 27, 34
Passive immunity, 142, 145, 148–151
Pathogen-associated molecular patterns (PAMPS), 24
PedvaxHIB, 26, 28
Peptidoglycans, 20
Peritoneal cavity (PerC), 106, 108–112, 114
Plasma cells, 106, 112, 121, 123
Pneumococcal capsule, 31
Pneumovax 23, 26, 31
Polyribosyl ribitol phosphate (PRP), 25, 28
Polysaccharide capsule, 18, 29, 34
Polysaccharides, 18, 20, 21, 23, 24, 27, 30, 31, 33, 106, 107, 113–118, 120, 121
Prevnar, 26, 32
Protection, 106, 109, 110, 112, 114, 121–123
PS-specific B cells, 21, 23

R

Relapsing fever, 63–70, 72–76, 78–80, 83–86

S

Secretory IgA, 138, 139, 144
Self-unresponsiveness, 5, 11
Serodiagnosis, 77–79
Sphingosine I-phosphate receptor 1 expression, 54
StaphVAX, 27, 32–34
Staphylococcus Aureus, 32–35
Streptococcus mutans, 133, 135–137, 140–151
Streptococcus pneumoniae, 20, 25–27, 31–32, 51, 56, 107, 112–115, 119
Streptorix, 27, 32

T

T cell dependent (TD), 21, 31, 69, 106
T cell-independent (TI), 18, 30
    antigens, 19, 25–35
TI-1, 106, 107, 116
TI-2, 106, 107, 113–120

T-independent response, 44
TLR2, 24, 28
TLR3, 52
TLR7, 52
TNF-α, 24
Toll like receptors (TLR), 2, 3, 9, 18–20, 24, 25, 106, 107, 109, 116, 118–120, 123
TriHIBit, 28
Type 1 TI antigens, 19, 20
Type 2 TI antigens, 19
Type b capsular polysaccharide, 28
Type I IFN, 52
Typherix, 26, 35
Typhim Vi, 26, 35

V
Vaccination, 2, 7, 10, 105, 106
Vesicular Stomatitis Virus (VSV-G), 3, 4
VH, 115, 116, 121, 123

W
Wall teichoic acid (WTA), 35

X
X-linked agammaglobulinemia (XLA), 106, 116
X-linked immune defect, 20
X-linked immunodeficient (XID), 106, 110–112, 115–120

Current Topics in Microbiology and Immunology

Volumes published since 1989

Vol. 271: **Koehler, Theresa M. (Ed.):**
Anthrax. 2002. 14 figs. X, 169 pp.
ISBN 3-540-43497-6

Vol. 272: **Doerfler, Walter; Böhm, Petra (Eds.):** Adenoviruses: Model and Vectors in Virus-Host Interactions. Virion and Structure, Viral Replication, Host Cell Interactions. 2003. 63 figs., approx. 280 pp.
ISBN 3-540-00154-9

Vol. 273: **Doerfler, Walter; Böhm, Petra (Eds.):** Adenoviruses: Model and Vectors in VirusHost Interactions. Immune System, Oncogenesis, Gene Therapy. 2004. 35 figs., approx. 280 pp. ISBN 3-540-06851-1

Vol. 274: **Workman, Jerry L. (Ed.):** Protein Complexes that Modify Chromatin. 2003. 38 figs., XII, 296 pp. ISBN 3-540-44208-1

Vol. 275: **Fan, Hung (Ed.):** Jaagsiekte Sheep Retrovirus and Lung Cancer. 2003. 63 figs., XII, 252 pp. ISBN 3-540-44096-3

Vol. 276: **Steinkasserer, Alexander (Ed.):** Dendritic Cells and Virus Infection. 2003. 24 figs., X, 296 pp. ISBN 3-540-44290-1

Vol. 277: **Rethwilm, Axel (Ed.):** Foamy Viruses. 2003. 40 figs., X, 214 pp.
ISBN 3-540-44388-6

Vol. 278: **Salomon, Daniel R.; Wilson, Carolyn (Eds.):** Xenotransplantation. 2003. 22 figs., IX, 254 pp. ISBN 3-540-00210-3

Vol. 279: **Thomas, George; Sabatini, David; Hall, Michael N. (Eds.):** TOR. 2004. 49 figs., X, 364 pp. ISBN 3-540-00534X

Vol. 280: **Heber-Katz, Ellen (Ed.):**
Regeneration: Stem Cells and Beyond. 2004. 42 figs., XII, 194 pp. ISBN 3-540-02238-4

Vol. 281: **Young, John A. T. (Ed.):** Cellular Factors Involved in Early Steps of Retroviral Replication. 2003. 21 figs., IX, 240 pp.
ISBN 3-540-00844-6

Vol. 282: **Stenmark, Harald (Ed.):**
Phosphoinositides in Subcellular Targeting and Enzyme Activation. 2003. 20 figs., X, 210 pp. ISBN 3-540-00950-7

Vol. 283: **Kawaoka, Yoshihiro (Ed.):**
Biology of Negative Strand RNA Viruses: The Power of Reverse Genetics. 2004. 24 figs., IX, 350 pp. ISBN 3-540-40661-1

Vol. 284: **Harris, David (Ed.):** Mad Cow Disease and Related Spongiform Encephalopathies. 2004. 34 figs., IX, 219 pp. ISBN 3-540-20107-6

Vol. 285: **Marsh, Mark (Ed.):** Membrane Trafficking in Viral Replication. 2004. 19 figs., IX, 259 pp. ISBN 3-540-21430-5

Vol. 286: **Madshus, Inger H. (Ed.):** Signalling from Internalized Growth Factor Receptors. 2004. 19 figs., IX, 187 pp. ISBN 3-540-21038-5

Vol. 287: **Enjuanes, Luis (Ed.):** Coronavirus Replication and Reverse Genetics. 2005. 49 figs., XI, 257 pp. ISBN 3-540- 21494-1

Vol. 288: **Mahy, Brain W. J. (Ed.):** Foot-and-Mouth-Disease Virus. 2005. 16 figs., IX, 178 pp. ISBN 3-540-22419X

Vol. 289: **Griffin, Diane E. (Ed.):** Role of Apoptosis in Infection. 2005. 40 figs., IX, 294 pp. ISBN 3-540-23006-8

Vol. 290: **Singh, Harinder; Grosschedl, Rudolf (Eds.):** Molecular Analysis of B Lymphocyte Development and Activation. 2005. 28 figs., XI, 255 pp. ISBN 3-540-23090-4

Vol. 291: **Boquet, Patrice; Lemichez Emmanuel (Eds.):** Bacterial Virulence Factors and Rho GTPases. 2005. 28 figs., IX, 196 pp. ISBN 3-540-23865-4

Vol. 292: **Fu, Zhen F. (Ed.):** The World of Rhabdoviruses. 2005. 27 figs., X, 210 pp.
ISBN 3-540-24011-X

Vol. 293: **Kyewski, Bruno; Suri-Payer, Elisabeth (Eds.):** CD4+CD25+ Regulatory T Cells: Origin, Function and Therapeutic Potential. 2005. 22 figs., XII, 332 pp.
ISBN 3-540-24444-1

Vol. 294: **Caligaris-Cappio, Federico, Dalla Favera, Ricardo (Eds.):** Chronic Lymphocytic Leukemia. 2005. 25 figs., VIII, 187 pp. ISBN 3-540-25279-7

Vol. 295: **Sullivan, David J.; Krishna Sanjeew (Eds.):** Malaria: Drugs, Disease and Post-genomic Biology. 2005. 40 figs., XI, 446 pp. ISBN 3-540-25363-7

Vol. 296: **Oldstone, Michael B. A. (Ed.):** Molecular Mimicry: Infection Induced Autoimmune Disease. 2005. 28 figs., VIII, 167 pp. ISBN 3-540-25597-4

Vol. 297: **Langhorne, Jean (Ed.):** Immunology and Immunopathogenesis of Malaria. 2005. 8 figs., XII, 236 pp. ISBN 3-540-25718-7

Vol. 298: **Vivier, Eric; Colonna, Marco (Eds.):** Immunobiology of Natural Killer Cell Receptors. 2005. 27 figs., VIII, 286 pp. ISBN 3-540-26083-8

Vol. 299: **Domingo, Esteban (Ed.):** Quasispecies: Concept and Implications. 2006. 44 figs., XII, 401 pp. ISBN 3-540-26395-0

Vol. 300: **Wiertz, Emmanuel J.H.J.; Kikkert, Marjolein (Eds.):** Dislocation and Degradation of Proteins from the Endoplasmic Reticulum. 2006. 19 figs., VIII, 168 pp. ISBN 3-540-28006-5

Vol. 301: **Doerfler, Walter; Böhm, Petra (Eds.):** DNA Methylation: Basic Mechanisms. 2006. 24 figs., VIII, 324 pp. ISBN 3-540-29114-8

Vol. 302: **Robert N. Eisenman (Ed.):** The Myc/Max/Mad Transcription Factor Network. 2006. 28 figs., XII, 278 pp. ISBN 3-540-23968-5

Vol. 303: **Thomas E. Lane (Ed.):** Chemokines and Viral Infection. 2006. 14 figs. XII, 154 pp. ISBN 3-540-29207-1

Vol. 304: **Stanley A. Plotkin (Ed.):** Mass Vaccination: Global Aspects – Progress and Obstacles. 2006. 40 figs. X, 270 pp. ISBN 3-540-29382-5

Vol. 305: **Radbruch, Andreas; Lipsky, Peter E. (Eds.):** Current Concepts in Autoimmunity. 2006. 29 figs. IIX, 276 pp. ISBN 3-540-29713-8

Vol. 306: **William M. Shafer (Ed.):** Antimicrobial Peptides and Human Disease. 2006. 12 figs. XII, 262 pp. ISBN 3-540-29915-7

Vol. 307: **John L. Casey (Ed.):** Hepatitis Delta Virus. 2006. 22 figs. XII, 228 pp. ISBN 3-540-29801-0

Vol. 308: **Honjo, Tasuku; Melchers, Fritz (Eds.):** Gut-Associated Lymphoid Tissues. 2006. 24 figs. XII, 204 pp. ISBN 3-540-30656-0

Vol. 309: **Polly Roy (Ed.):** Reoviruses: Entry, Assembly and Morphogenesis. 2006. 43 figs. XX, 261 pp. ISBN 3-540-30772-9

Vol. 310: **Doerfler, Walter; Böhm, Petra (Eds.):** DNA Methylation: Development, Genetic Disease and Cancer. 2006. 25 figs. X, 284 pp. ISBN 3-540-31180-7

Vol. 311: **Pulendran, Bali; Ahmed, Rafi (Eds.):** From Innate Immunity to Immunological Memory. 2006. 13 figs. X, 177 pp. ISBN 3-540-32635-9

Vol. 312: **Boshoff, Chris; Weiss, Robin A. (Eds.):** Kaposi Sarcoma Herpesvirus: New Perspectives. 2006. 29 figs. XVI, 330 pp. ISBN 3-540-34343-1

Vol. 313: **Pandolfi, Pier P.; Vogt, Peter K. (Eds.):** Acute Promyelocytic Leukemia. 2007. 16 figs. VIII, 273 pp. ISBN 3-540-34592-2

Vol. 314: **Moody, Branch D. (Ed.):** T Cell Activation by CD1 and Lipid Antigens, 2007, 25 figs. VIII, 348 pp. ISBN 978-3-540-69510-3

Vol. 315: **Childs, James, E.; Mackenzie, John S.; Richt, Jürgen A. (Eds.):** Wildlife and Emerging Zoonotic Diseases: The Biology, Circumstances and Consequences of Cross-Species Transmission. 2007. 49 figs. VII, 524 pp. ISBN 978-3-540-70961-9

Vol. 316: **Pitha, Paula M. (Ed.):** Interferon: The 50th Anniversary. 2007. VII, 391 pp. ISBN 978-3-540-71328-9

Vol. 317: **Dessain, Scott K. (Ed.):** Human Antibody Therapeutics for Viral Disease. 2007. XI, 202 pp. ISBN 978-3-540-72144-4

Vol. 318: **Rodriguez, Moses (Ed.):** Advances in Multiple Sclerosis and Experimental Demyelinating Diseases. 2008. XIV, 360 pp. ISBN 978-3-540-73679-9